Zoology
Believe It Or Not!

We would like to welcome you to the wonderful world of zoology. We have a diverse community of animals worldwide. There is wonder and mystery surrounding us that can be easily taken for granted. We would like to welcome you to our world of animals both real and mythical. Can you tell the difference between the real and mythical animals in this book?

Written and Illustrated in the fall of 2013 at High Tech Middle School in San Diego by the 6th Grade Students of Maria Schaffer and Ben Krueger.

We would like to thank our special 4th grade friends at Explorer Elementary for your time, effort, and consultation to help us make a better coloring book.

<u>**Real and Mythical Key:**</u> All individual writing pieces and illustrations are real animals, and all partner written pieces and illustrations are mythical animals combining aspects of both real animals.

Red Panda
By Sierra Gillingham

"What do I want to eat this morning? Fruit or bamboo? Lichens or acorns? Insects or eggs? Maybe if I get lucky I can find some berries or mushrooms." I thought to myself.

I ended up finding some berries and bamboo. I used my false thumb to help me eat.

When I was all full, I headed back home to my tree. My tree was a little bit aways from all the tasty food, but I like my tree because it was in the cool parts of China. When I got back to my tree I climbed up and layed down with all my four legs hanging off the side of the branch. I look around. I see another red panda that was a little bit aways, since we are solitary animals, we never interact. It was dark out because I only like to be awake at dusk and dawn. I close my eyes and wait for it to turn to dusk.

When I wake up it's dusk and the sun is going down. I get up and climb down my tree. I don't have to worry about another red panda coming on to my tree because I mark my territory by the pads on my feet that leave a scent wherever I walk. I leave my tree and go look for more food. I'm in the mode for some fruit.

While I'm out on my hunt I spot one of my three predators, the snow leopard. I hide in a tree that has the moss I blend in with. The moss is the red color of my fur. I stay completely still. If the snow leopard had seen me and came chasing at me I would have stood on my hind legs to make me look taller and made hissing noises to scare him. I'm glad the snow leopard didn't get me because I am an endangered species. As the snow leopard moves on, I slowly come down the tree. I continue my hunt. I spot something.

As I was eating my prey I saw another red panda. He was walking by and ignored me. I watched him walk off.

"I wonder when I will be ready to mate. I wonder if I would have 1, 2, 3, or 4 cubs? Then they can play, wrestling, bite, and lung. I hope they wouldn't hurt each other." I thought to myself. By the time I finish my feast it was already the middle of the night. I head back to my tree.

When I get back to my branch I use my fluffy tail that has a ring pattern on it to cover my 24 inch long body that weighs 15 pounds. I do this when it is cold out because my tail is so warm, but if it is warm I don't need my tail. I also use my tail for balance when I am in trees. I felt like I had dirt and moss all over me so I used my tongue and licked every part with dirt and moss.

After I was all clean, I closed my eyes. It was a little bit before sunrise so I fell asleep. I was full and sleepy. I wonder what I'll want for my next meal? Eggs or fruit? Bamboo or insects? Acorns or mushrooms? Maybe some berries. I drift off to sleep feeling good about being a red panda.

Blue Jay
Julie Rodriguez

As the hawk calls out loud, my heart is pounding with fear. I can hear the hawk coming closer every minute. I start in and jump back against the tree until I can't move back anymore. It starts to circle around the tree. I start to imitate a hawk. It flies away seconds later. I am so lucky to be standing on my two feet right now. I look outside from the tree to see if everything is clear and I fly away. The sun reflects off my feathers turning my feathers a brighter blue. I also live in a flock of 250 other blue jays, which are really good at hunting and taught me well. My classification is Aves.

Everyday my sister, Eva and I go hunt for food together because we work good as a team. We like to hunt mice, frogs, insects, and at times we take other birds eggs, but we also eat fruit, seeds, nuts, and acorns. We are omnivores, but it really doesn't matter. I like to hide acorns and seeds in the ground because if we start to run out of food, we can just dig it up and enjoy. We have to avoid larger birds like owls, falcons, and hawks.

I don't grow as much as other blue jays. My wingspan is about 13 to 17 inches long. I can live up to 7 years or more. I don't weigh very much. I weigh about 70 or 100 grams or 2.5 and 3.5 ounces. I'm around 22 to 30 cm. tall. I'm smaller than an owl. I have bright blue feathers on my tail and back. I also have a lot of black on my throat, feet, beak, legs, eyes, and head. My face has a striped pattern colored black and white. My chest is black and white. I can tell the difference between a male and a female blue jay because a female is smaller than a male bird. My scientific name is, Cyanocitta Cristata. Sometimes I can be really noisy to warn other blue jays that predators are nearby and are ready to hunt. It gives them time to hide in trees.

My flock moves a lot in different seasons. We will be breeding up in the north during the summer. In the winter, we are in a small portion of the southwest when we aren't breeding for the season. But year round, we are mostly in the east or west. We also can be found in oak trees in mixed forests because of the selection of food. I can be found in the states of Florida and Texas and the country of Canada.

I wonder what it's like to reproduce. I only know somethings about it, like a bird can lay 3 to 6 eggs. They can be different colors like blue, green, and yellow with grey or brown spots. About 1 year after a male or female has been born, they can start to reproduce. They can stay with their partner until one of them dies. They can make their nest out of twigs, leaves, grasses for trees, and shrubs. When their nest is complete, it is fragile because it is loosely put together. They reproduce in the months of march through july.

I like to be a blue jay even though it is harder than it seems. Trying to dodge our predators and look for food. I'm glad to be a blue jay. I come with awesome features and a beautiful shade of blue. A challenge is to get places by not being afraid of my surroundings and also by helping other blue jays.

Janda

By Sierra Gillingham and Julie Rodriquez

    I was flying through the sky migrating north for the summer, with my purple, white, and black wings keeping me up in the air. My purple fur was blowing in the wind and I was feeling free. I was flying with my flock of 250 other Jandas. My sister was flying next to me as we were trying to find food to satisfy our midnight hunger. We like to hunt at dusk and dawn because my prey comes out at night and my predators are asleep. I sometimes eat little rodents like mice who come out at night and I also eat plants, bamboo, berries, acorns, fruit, mushrooms, and insects. Once or twice I had bird eggs. I hear something in the bushes...

    I saw my predator, the snow leopard. I call out to all the other Jandas to warn them about the snow leopard. Then, when the snow leopard sees me I stand on my hind legs and make hissing and snorting sounds so it will make me look bigger than I was before and that I will appear scary and frightening. As the snow leopard leaves, I begin to get really hungry.

    I spot some berries in the grass a few feet away on a branch of a bush. I walk over to the bush and I use my thumb that is called a "panda's thumb", to grasp the branch and eat it. The berries are delicious! Once I am done, I head back to my tree.

    It was late and I wouldn't have made it back to my tree before morning. Morning is dangerous because it is when most of my predators come out, so I had to move fast to find another tree. Fortunately, the mountains in China have lots of trees that are perfect for me so I shouldn't have to worry. If I lived in western Texas or southern Canada where some of the other flocks live, it would have been tougher to find another tree. A worry I have for finding a tree, is finding an oak tree because the trees give protection, safety, and they are strong so they can hold my entire flock.

    Finally, I find a tree that can hold my weight, 14 pounds. The branch is 45 inches long and I am 23 inches long so it is a good fit. I spread myself out on the branch to find the right spot. Then I curl my ring patterned tail around my body because it is cold out and my tail i so warm that it could warm my entire body up. After a while, I feel dirty after all the adventures I had today. I get up and use my tongue to lick myself clean. While I am cleaning myself, I use my tail to balance in the tree so I won't fall.

    When I am done cleaning, it is already morning and I had to go to sleep, so I will be wide awake for tonight. As I am falling asleep, I can hear two other jandas communicating by squeaks and whistles for a few minutes to warn the other jandas that there is a hawk in the area. I close my eyes and drift off to sleep.

    As I am sleeping, I have a dream. I am looking for food but it isn't in my territory. I hope no one goes in my territory while I am lost because I am a territorial animal. I start to fly up in the air so I can look around to see if my home was in sight but it is pitch black. I look down for food and spot a white dot on the ground. In curiosity, I fly down, but using my intelligence I make sure not to scare it because if it is prey it could run away. I get closer and I see it is a mouse. Two trees appear. The mouse runs between the two trees. My wingspan is 15 inches long and the trees are about 17 inches apart, so I am able to fly between them. As I am really close to getting my prey, I suddenly wake up to the sounds of a janda squeaking for help. I race down my tree and go help the other janda.

    I as I arrived to the other janda, I see a hawk trying to attack the janda. As I approached the hawk, I stand on my hind legs and start making hissing noises so it would scare the hawk away. It works, it scares the hawk away and it never comes back. I head back to my tree feeling proud for saving that janda.

    When I get back to my tree, it is dawn. I feel good about being a janda even though it can have tough challenges, it can also be really fun to hunt, fly, and play with my sister. I love being a janda because I get to be free.

Echidna
By: Riley Wells

Rummaging around in the great victorian desert of Australia, is a small-spiny creature. It is about 14-30 inches long and has a brown fur color, which was great for camouflaging. What was it? Reminds me of an anteater, a porcupine, and a hedgehog. My studies have failed me and I found something completely different than my first idea. I sat down at my desk in the makeshift hut. I wrote in the journal:

"It was a female. She was incredible. She was accompanied by a juvenile. They have never been discovered before. I looked through mountains of books and nothing about this strange animal."

I went back out to that same spot I found the mother and her baby. She was still there. I decided to call her species an echidna. I am greek, and so I named the species after a mythological monster that was originated in my home country.

I ran to my truck and drove to the mayor of the city I was in and got permission to own and study the creature I call the echidna. So I rushed back to the hut again.

I studied a male and a female. I ran some tests and I saw they eat insects, earthworms, ants, and other types of bugs. The male and female stuck out their long-slim tongues and caught their meals. I also discovered that they communicate by smell! The echidnas can sense each other's feelings. It is kind of like dogs. Dogs can also sense feelings. I named my echidnas, Ta-Ta (the male) and Rica (the female.) I took them all over the world. They can live in any climate! I found other echidnas in Tasmania, Indonesia, Australia, and Papua New Guinea! They adapted so well that I could not believe my eyes.

On my journey around the world with Ta-Ta and Rica, I saw that most echidnas used their spiny backs as defense whenever a dingo, eagle, or big cat came near. The animal wouldn't even try to hurt the little guys. I know I wouldn't! These animal must be their predators.

When Ta-Ta and Rica became of age they mated and Rica laid a single egg and it hatched in 10 days. I named the puggle, pugs because of his cute little face. I decided to call baby echidnas puggles. Pugs was born without spines and he stayed in Rica's pouch for about 55 days. I studied and saw that juvenile echidnas develop in the egg, but not from a yolk.

Ta-Ta and Rica finally died after the 16 years I had them. I was so sad, but atleast I still have Puggs. But soon, I let Puggs free back where I found Ta-Ta and Rica. Soon I went back to the same spot where I let Puggs free, again. I recognized him from his unique fur color. It made me so happy to see my old friend again, and what do you know Puggs started his own family. As he was leaving the newly started family because they are solitary creatures, II saw his partner. I was very proud of him for making it to the wild.

Great White Shark
Trajan Vavra

Great white sharks are built for speed and they are examples of the perfect predator. The oldest human remains that have been discovered are 1.3 million years old while the oldest great white fossils are about 16 million years old. Their bodies have hardly evolved at all in this huge timespan.

They are torpedo shaped which helps them slice through the water. Their Dorsal fin stops them from rolling and helps them make sharp quick turns. Their mouths can house up to 300 teeth and they can swim up to 40 mph. Their body is colored gray (to blend in with the rocky sea floor and their undersides are white to blend in with the sky. They can weigh up to 5,000 lbs. Great white sharks can grow up to 26ft but their average size is about 13 feet.

Great white sharks, scientifically known as Carcharodon Carcharias, sometimes travel over entire oceans in search of prey. They are found in the sunlight zone and the twilight zone(surface-820ft below sea level). Great whites prefer temperate water but live in about every ocean. Great whites have the longest recorded migrations of any fish species; from South Africa to Australia and from California to Hawaii! Great whites are able to live in the open ocean or around islands and continental coasts. They are able to live in frigid and warm water.

Great white sharks prey on many different types of sea life. Younger sharks eat fish and rays, while adult great whites eat other sharks, seals sea lions and some other sea mammals.This massive carnivore often approaches their prey from underneath them and sprints up and thrust their jaws up and takes a huge bite out of their target. Great white sharks rip up their prey and swallow chunks whole. Since great whites don't have eyelids, when they bite down, they roll their eyes back in their sockets. When great white sharks are hunting seals, they can breech up to 10ft, which means they can jump 10ft out of the water. Like a gray whale, this animal also checks their surroundings by spyhopping in search of prey. Spyhopping is when they jump straight up out of the water and look at their surroundings.

Although great whites are often thought of as the ocean's deadliest predators, these dangerous sharks actually fall prey to humans. Every year approximately 100 million sharks are killed. People use their bodies to make shark fin soup, shark liver oil capsules, skin cream, lipstick, and jewelry. Great whites conservation status is "vulnerable". Humans are also responsible for the destruction of their near-shore habitats, where their young are born and raised.

Contrary to what most humans believe, mankind is more of a threat to the great white species than these powerful sea creatures are to humans.

The Great White Esharkna

Trajan Vavra and Riley Wells

Esharknas are the most dangerous creatures on Earth. With their body covered in spines and their ability to stay in or out of water, they are truly at the top of the food chain.

The Great White Esharkna can live in many different oceans and climates. Surprisingly Esharknas have gills and lungs but only stay on land for about an hour due to water adapted eyesight. Esharknas live mostly in Australia and are concentrated in the Great Barrier Reef, but they also live in many other places like off the coast of Monterey and Oregon. Esharknas are tracked by professional marine biologists so they can see where they are because people fear them, and want to learn more about their behavior.

Great White Esharknas are giant fish/mammals. Their backs are grey, so from above they look like the sea floor and their stomachs are white, so from behind they blend in with the sky. They also have five claws on both of their front fins. The Great White Esharkna's mouth can house up to 300 gnarly teeth! The teeth are used for ripping its prey into pieces to then swallow instead of chewing. The Esharkna's dorsal fin helps it stop rolling and slice through the water. The Esharkna is free to swim up to 40 mph. Esharknas also have sharp spines on their backs which are used for defense and hunting.

Younger Great White Esharknas eat Terottermites while adults eat Sea Ants. These massive creatures will eat up to 500 lbs. on a daily basis. Esharknas are also cannibals and eat each other. An adult dominant male will eat juveniles or end up killing one in a fight. Other Esharknas are their only predator besides humans. Esharknas will occasionally eat the average human and is known to once and a while capsize a small boat. Esharknas have gills and lungs, but since they can't see on land, they only stay there for about an hour at a time.

Esharknas have very complicated lifestyles. Pupples, baby Esharknas either hatch out of an egg or they are birthed. It all depends on where they live. If they are in frigid water, they will lay an egg. If they are in warm water then they will give birth. When they are born, they are already 5 ft long, but if they hatch out of an egg they will only be 2 ft long. Pupples have to be extra careful because adults might only see them as prey. That is one reason why Esharknas are so protective. They learn to stay on guard from the beginning of their life.

Over all, Esharknas are very dangerous and bloodthirsty creatures. Watch out next time you're at the beach!

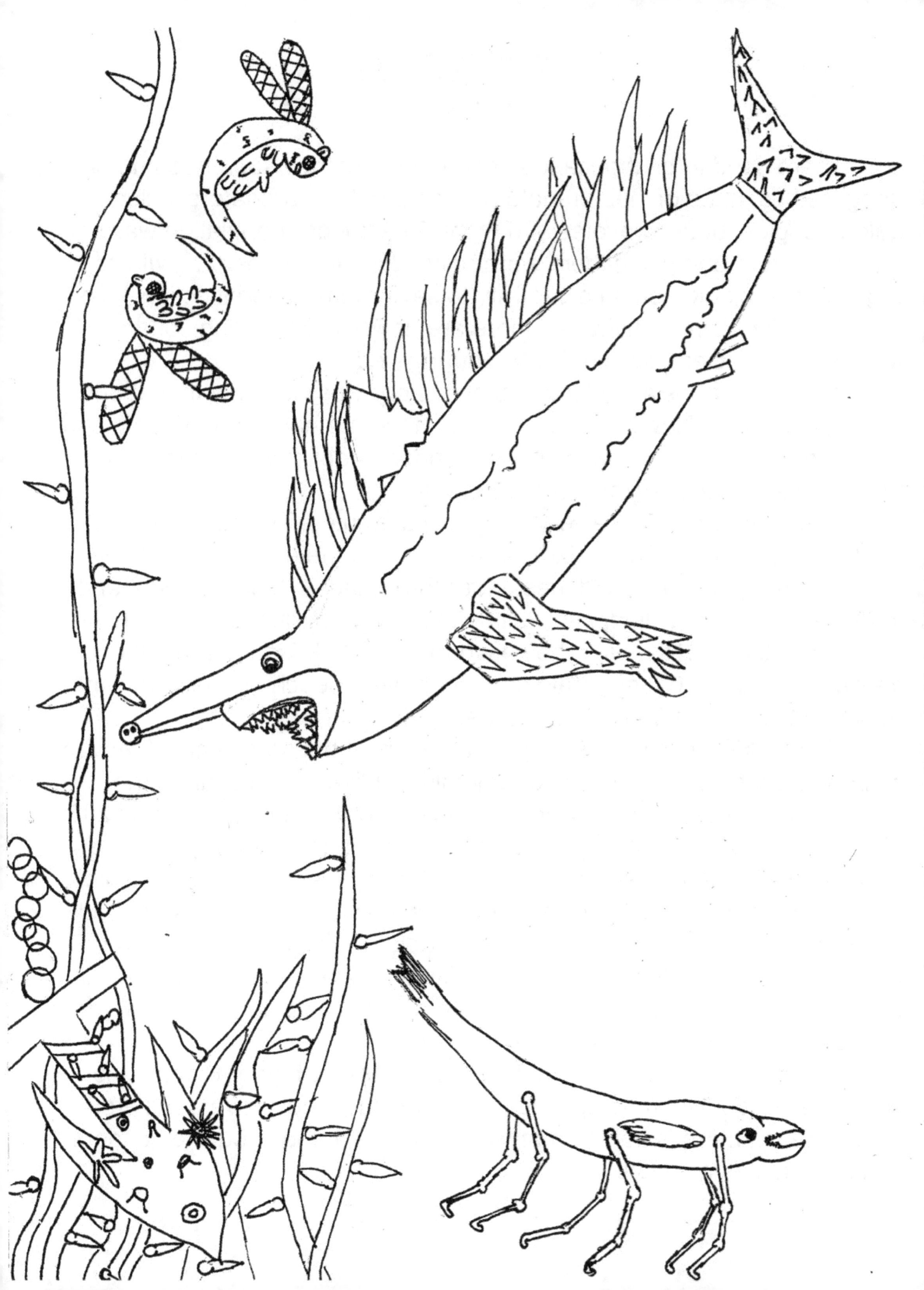

Condors
Liam Nolan

There used to be thousands of condors, but they soon became endangered because of their predators. Some of their predators include, golden eagles, bears, coyotes and more. The condor population was so endangered at one point, there were only 22 left on the earth. WIldlife organizations took them into captivity to save them from extinction. They are no longer all in captivity, but there are now 230 free flying condors today. The species however is still endangered.

The condor has a very interesting history. Fossils show that at one point they lived in New York and Florida. To some native american tribes, the condor is a symbol of power. The condor's habitat is the desert and they currently live in Arizona and California.

The condor has some very intriguing attributes. They are one of the world's largest birds. Their wingspan is almost 3 meters long and they have big triangular white patches under their wings. Their eye-sight and intelligence is extraordinary as well as their flying abilities. They can fly up to 15,000 feet in the air. They have mostly black feathers, but the juvenile's feathers are gray. Male condors have a pouch on their neck that they use to attract a female during mating season.

The condor mostly eats large dead animals but sometimes if they're really hungry they will eat smaller animals. Unlike most vultures, the young won't kill each other for food. The population has gone up by 208 condors in about 33 years. This is amazing because they only lay one egg every year so they don't have to risk having many of the chicks die.

Bears, eagles and coyotes weren't the only predators condors had to worry about. Hunters were another problem. They would use lead ammunition, to give the condor lead poisoning if the bullet itself didn't kill them. Therefore as long as the condor was hit, it was domed. Now, that they are endangered, it's illegal to hunt them.

There is only one other bird in the world that is bigger than the condor. The condor is a rare sighting to see because they have a big area of existence but there is not too many of them left. I hope you like condors just as much as me now.

Spotted Wobbegong
Jackson Ducksworth

200 feet below the surface of the ocean lies a shark. This shark is very peculiar, it is flat and has rounded fins. This shark is the spotted wobbegong. spotted wobbegongs have very cryptic colors, these colors include yellow, brown, and sometimes green. The patterns on the skin are usually O-shaped blotches and sometimes look like someone poured a bucket of paint on them.

The males are bigger and can grow to a length of ten and a half feet, females tend to grow to about eight feet. Spotted wobbegongs also have an interesting feature, they have barbels that hang down from around their snout. This interesting creature can also have six to ten lobes of skin that hang over their eyes. This shark may seem peaceful but when disturbed, it uses its skills of survival and can deliver a very powerful bite. Back to the barbels they use these "whiskers" to taste and feel. The shark may seem bulky and slow but when needed it can go lightning fast to escape predators or catch prey. When trying to catch their prey they can use their colors to hide among faded coral reefs. All these skills come in handy when hunting prey.

The spotted wobbegong is a very hungry shark and tries to catch prey no matter what. They feast on many invertebrates such as crabs, lobsters, octopi and also vertebrates that include sea bass and ludericks. They are nocturnal but sometimes they hunt in the day. When hunting they lay on the seabed and cover their body with sand. Here they wait for prey to wander close then they expand their throats and create a powerful vacuum that sucks in the prey. Their large mouths aid them in this process. They don't have many predators but, bigger fish and large mammals are potential predators.

The spotted wobbegong can be found in only one place and that is off the coast of eastern australia. They can be found down to 360 ft., this is also known as the sunlight zone, on sandy seabeds and can also be found under piers, docks, coral reefs, and rocky reefs. Scientists also believe that they inhabit the southern china Sea and also can be found off the coast of japan. They may also be found in warm waters.

This beautiful and majestic creature is one that not many people know about. They are somewhat endangered and need to be kept alive so the people of the world can notice them and care for them. So notice this wonderful creature and make sure it stays alive forever.

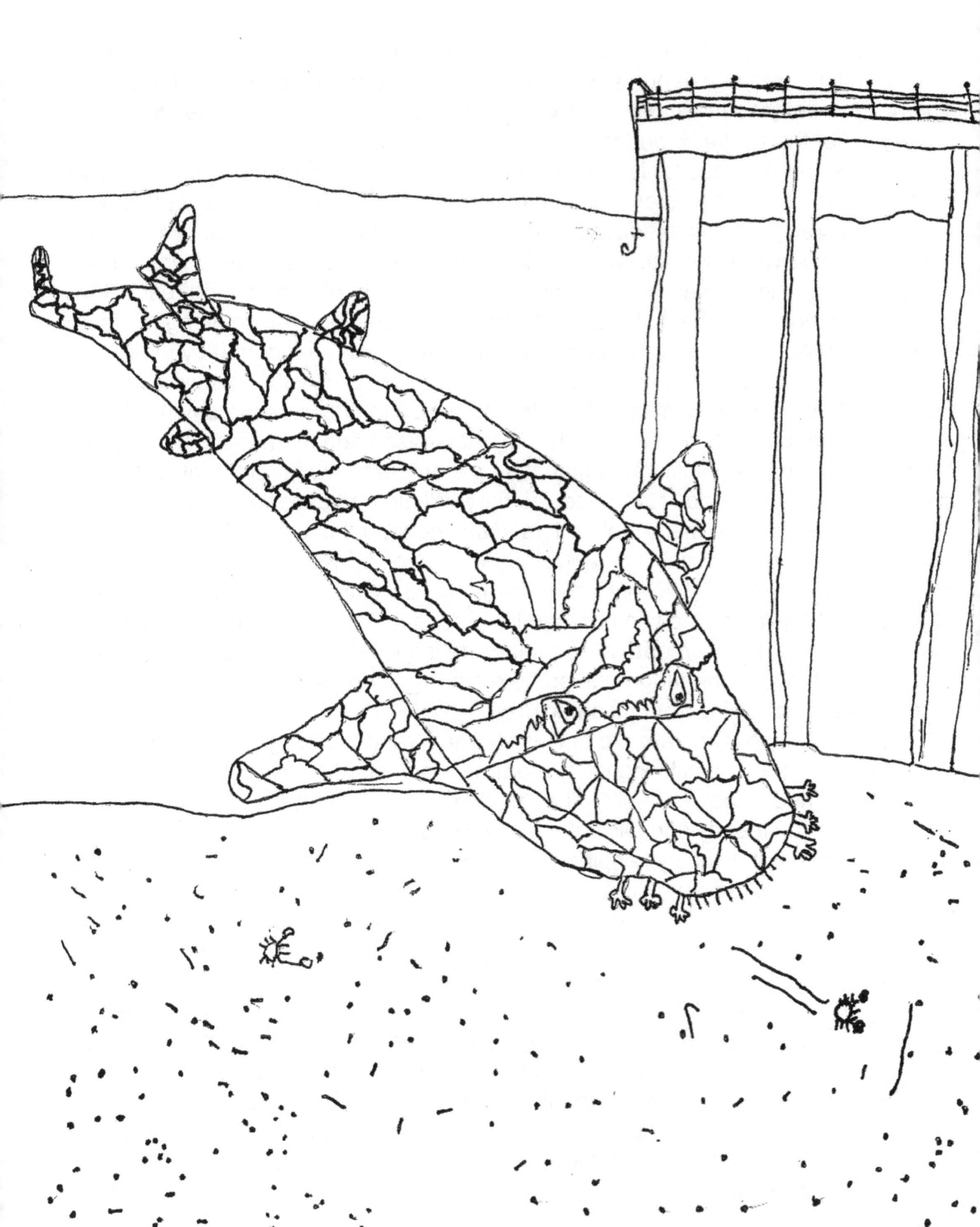

The Spotted Wobbedor Vulture of the Sea

Liam Nolan and Jackson Ducksworth

    The spotted wobbedor is a creature many people will not see in a lifetime. It is very endangered due to humans and overpopulation of mammals and large fish. The habitat, defense mechanisms, hunting, and young are all part of the world of the hungry life of the spotted wobbedor.

    The spotted wobbedor is a ruthless animal that is classified as a fish. They will do anything to catch their prey. They have pectoral fins that are specially adapted to let them crawl across the seabed. They also have large mouths that can suck in prey when a powerful sucking motion is created. This motion is created when the sharks expand their throat. They are nocturnal so they hunt at night and rest in the day. The wobbedor is the vulture of the sea, this means they are scavengers as well as hunters. They search for dead fish that have been abandoned by their original predators. Wobbedors have tough skin that it uses to withstand blows from prey and predators. They have nerves in their feathers so they can sense any predators that are trying to sneak up on them. The feathers also keep them warm so they can hunt in deeper water. Most of the time they are on the move for prey unless it is mating season. If they don't keep moving they will die and become extinct.

    There used to be many wobbedors in the world, but because they have been hunted they are now critically endangered. At one point ,there were only twenty nine wobbedors left in the wild. People from wildlife conservations have increased the population of this animal to two hundred thirty six but they are still trying to raise it. The wobbedor has few predators, some include large fish, mammals, and man itself. Some of their more dangerous predators are golden eagles and bears. With its cryptic colors it can hide from these animals.

    The wobbedor compared to man is a large strange looking creature. They have a broad flat body and when fully matured have a body length of ten and a half feet. The spotted wobbedor has O-shaped blotches covering their skin and feathers that cover their body and allow them to swim deeper than normal. The feathers help insulate their bodies by keeping them warm. They come in yellow, brown, and green, and sometimes are all three. Their feather colors come in gray and black. They have a wingspan of almost three meters, about seven feet for those of you not familiar with the metric system. When born their young have a wingspan of three feet and a length of five feet.

    Spotted wobbedors have large families. They lay one egg each year that hold up to thirty seven wobbedor pups. Many of these pups don't survive which is another reason why they are so critically endangered. Like all sharks, the young are born out of egg sacs. When they are born, they already know how to hunt and survive. Now that the world of the spotted wobbedor is explained people will make sure to save them and not hunt them anymore. Wildlife conservations have saved them and they will continue the legacy.

Leafy Sea Dragon
By Alexis Covey

The leafy sea dragon might have an intimidating name, but it is a harmless fish with very few defense mechanisms. Its beautiful leafy appendages can be seen flowing in shallow waters. It looks calm and relaxed compared to some of the other creatures that can be found in the ocean. In order to survive, they use their many different adaptations.

Their main survival skill is their amazing camouflage. Their entire body is covered with colorful leaf like appendages and tiny plates acting as armor. The leafy sea dragon if it is healthy and has enough food to give it energy, is able to change its body color to blend in with the environment. The most common colors of sea dragon are brown, yellow and olive. They are very similar looking to the seahorse. They grow up to 14 to 18 inches in length. Their life span is from 6 to 8 years.

They have a hard time eating with their snout because it is so long and narrow. The leafy sea dragon's favorite food is mysid shrimp. They also eat sea lice, plankton, and larval fish, but their main source of food is amphipods. The sea dragon constantly has to eat by sucking up their food because they have no stomachs or teeth.

Leafy sea dragons have very few predators. Instead of swimming upright like seahorses they swim headfirst. Their body armor slows them down when swimming, making them easy prey, but they are such experts in camouflage that they are very difficult to hunt. Only flounders are sometimes able to hunt sea dragons. The leafy sea dragon is also under the protection of Australian laws. They were almost pushed to extinction before due to hunters.

Sea dragons love swimming in the wonderful Pacific Ocean off the coast of Australia and in the eastern Indian Ocean. The areas in the nice waters they live around are kelp forests, rocky environments, seagrass meadows, and sand patches. The leafy sea dragon swims in waters from the surface down to 82 feet. The temperatures they like are from 55 to 67 degrees.

These animals are not normal when it comes to their youngs. In this strange species the male has the babies instead of the females. The male keeps the eggs on the outside of their tale. The eggs hatch from 6 to 8 weeks. When the eggs are ready to hatch the male shakes the eggs and the tiny baby sea dragons are born. They can have 250 eggs to 300. When they hatch they eat zooplankton and are 0.16 inches in diameter and 0.3 inches long. When the sea dragons are two they are fully grown.

The leafy sea dragon is an amazing creature. The males have the babies, they have the ability camouflage, and they are delicate animals. These fascinating animals, sadly are almost extinct and could soon be only legends like dragons.

Wolverine
Johnny Delgado

Wolverines are known to be strong and vicious. They are somewhat small and muscular. They are part of the weasel family, but they look like skunk bears. Even though they are small, they hunt animals that are bigger than them. They are tough and interesting creatures.

Wolverines have a lifespan of 7 to 15 years. They can be between 24 to 36 inches from their nose to their tail. Their tail alone can measure between 7 to 10 inches. A wolverine's weight varies anywhere between 24 to 40 pounds. They have extremely thick fur that can be used by humans as a coat.

The wolverine features and many different attributes allow it to survive. The wolverine has thick fur so it doesn't freeze. They also have a strong smelling fluid like the skunk that allows them to mark their territory. They have sharp teeth so they can kill their prey as well as having a rotated molar that allows them to rip flesh off of an animal that has been frozen all the way through. They also have powerful jaws so they can chew their prey. The wolverine has strong front legs with claws to rip up and hunt their prey. It also has wide feet to climb on trees and their feet are webbed so they can swim.

Wolverines live in Canada and Boreal forests. They live in Asian, European and North American tundras and taigas. Wolverines are mainly solitary animals. Female Wolverines burrow into the snow in february to create a den. They will keep this den until they give birth and their babies are ready to hunt.

Wolverines are strong enough to take down animals much larger than themselves, but they mostly eat smaller things like rabbits, mice, rats, birds and bird eggs. Wolverines sometimes go for bigger animals like Caribou and Lynx. In the summer, wolverines eat small animals, eggs, berries and other vegetation. In the winter, wolverines can also eat hares, grouses, deer, reindeer, and mooses.

Wolverines are small but fierce animals. They use their adaptations to hunt, build shelter, and protect themselves from predators. These creatures are one of the toughest small mammals in the animal kingdom.

The Leafy Swolverdragon
Johnny Delgado and Alexis Covey

The leafy swolverdragon is known for its super strong jaws and amazing leafy appendages. This amphibian is a master of camouflage and has one of the most unique appearances of any creature on land or sea.

The swolverdragon has the appearance of a swamp monster, with leafy appendages protruding from all over its body. On its skin where there not any appendages, it is covered in a greenish brown fur. The swolverdragon is 20 to 30 inches in length. It weighs about 10 to 15 pounds. They are stronger than they appear. They have sharp claws and teeth to overpower their prey. Their snout is long to help them sniff out their prey. Under the sea it can hold it's breath for about 3 hours. It's webbed paws help it to swim better. Its fur is thick to help it survive cold climates.

The Leafy swolverdragon lives in the Atlantic Ocean and in Europe, on land. Underwater it lives in seaweed patches and sand patches. When on land they live in snowy places with many trees. They are known to stay as far away from human populations as possible. They mostly live in water though they do like to go on land when hunting for prey.

They have a lot of prey to catch in the water and on land. The swolverdragon has it's long snout with sharp teeth to chew berries and catch prey. The swolverdragon is able to catch prey much larger than itself because of its strenghth. The animals they eat are caribou, sea lice, lynx, larval fish, plankton and hares. It searches for food at the bottom of the ocean, in trees and on land. In order to kill their prey they rip off their flesh and eat the animal.

Along with their sharp teeth and killer instinct, swolverdragons have many other unique survival skills. They also have to use their camouflage in the water to escape from predators. Surprisingly, the only known predator for the swolverdragon is the Flounder. These are the only fish that can see through the swolverdragon's disguise.

The leafy swolverdragon is an unlikely animal looks as though its out of a fairytale. Their rare appearance and elusive nature have made swolverdragons a very sought after creature.

The Arctic Fox
Liv Hicks

Arctic Foxes are most often thought of as cuddly and cute balls of fluffy fur, but in reality, they are resourceful, daring hunters designed to survive in very harsh conditions. Every way that they eat, sleep, or hunt is solely dedicated to living through anything.

In the barren Arctic Tundra, food is hard to find even in the summer. Arctic Foxes eat whenever they can, and during the summer they feast on other animal's young, such as bird eggs, or left carcases from larger predators like polar bears. They also collect berries when available. Arctic Foxes hunt small prey with their expert hearing. They slowly prowl the landscape listening for underground sounds of animals in their burrows. When they pick up a sound, they can locate the exact place where the animal is, and jump on top of the snow directly over the animal killing it quickly.

Arctic Foxes live in one of the most unforgiving habitats in the world. The Arctic Tundra is home to the Arctic Fox, it lives on the edge of it's forests. Several feet of snow cover the ground in the Winter. These conditions are so harsh because the snow is constantly raining down on the landscape, averagely 4-6 ft. As winter passes Spring brings out fresh berries and hunt. Snow slowly melts, but not much. Summer and fall go by without much change, and winter comes again.

Arctic Foxes look like deadly cuddlers. With beautiful thick white fur, their Winter coat keeps them warm even when temperatures reach -50 C. Arctic Foxes have the thickest coat in the Arctic. They have long snouts, and small black eyes. Their bodies are very close together to provide easy body heat. They are a little larger than the average house cat.

Arctic Foxes are social animals. They live in groups of a paired male and female, and a second female to care for the pups. She will not mate. Females are averagely pregnant for 50 to 60 days. They produce 9 to 14 pups. The Arctic Fox is very protective over it's territory and young. Both the mated male and female will defend their pups and burrow ficiously.

The Arctic Fox is incredibly armed against the tough habitat of the Tundra. Remember to learn about something before you judge it, as you have learned from the said "cuddly" Arctic Fox.

The Leatherbacks
Mickey Browne

Leatherbacks may seem cute and cuddly, but they have many features that make them different from other sea turtles. In fact, they are known as the sea giant. Their habitat, features, diet, and survival make them one of the most interesting turtles that live in the sea. Unfortunately, their species is dying out and could even be be extinct in 20 years.

Leatherback sea turtles are found all over the world. They are found in the Atlantic, Pacific, and Indian Oceans, but they are mostly found in the Atlantic Ocean off the coast of Europe and USA. They also have been found as north as Alaska and as south as Africa. Their widespread habitat is unique for the turtle species. Certain human actions like oil spills are threatening their homes. Their habitat provides them with a lot of food to eat.

Leatherbacks have an enjoyable menu of food to eat. They like to eat invertebrate sea creatures such as jellyfish because of its soft body. They also eat squid because it has the same texture of jellyfish. They have really good jaws for soft animals. They do eat crabs and other crustaceans with hard shells even though they are bad for their jaws. This is similar to kids eating too much candy, it is delicious, but not always good from them. Their jaw is only one of their important features.

The leatherback looks different than other sea turtles. They are the largest reptiles in the world. Their average height is 4-9 feet tall and a couple of them were even measured 10 feet tall. Their average weight is 660-1,100 pounds. From beak to tail it is 2,019 pounds. They can live between 40-45 years. This is short compared to sea turtles and one reason for it could be their size. Their size is not the only reason they have a short life span.

These turtles have hard lives when it comes to survival. 90% of animals in the Pacific Ocean have died because of the pollution like plastic in the ocean. Leatherbacks also sometimes mistake balloons for jellyfish. They have natural predators as well. Sea birds, for example, swoop down to eat their eggs before they are ready to hatch. This is particularly common during nesting season.

The leatherback is one of the most unique and amazing reptile in the oceans of the world. They can dive down 4,000 feet into the water from the sunlight zone part of the ocean where they live, they live all in many different oceans, and they are the only turtle that doesn't have a hard shell. It is important to keep our oceans clean to save these precious creatures.

The Foytile
Liv Hicks & Mickey Browne

You may not have heard of it, but the Foytile is a graceful and majestic amphibian that is often thought of as silly or unthreatening. It's massive body, sharp teeth, and powerful tail proves otherwise.

Foytiles have large habitats because they live in and out of the water, like most amphibians. They are found regularly in the Pacific Ocean, but also have been spotted in the Atlantic and Indian Oceans. All Foytiles migrate towards Alaska for the spring to arrive there in the summer. During this migration, they're found off both the west coast and east coast of the United States. In the cold snow of Alaska, they dig burrows to lay their eggs. Unfortunately, because of global warming their Alaskain home is melting and their food source is dwindling.

The Foytiles have many different food items that they hunt or find. When on their journey toward Alaska, they eat jellawata birds. Jellawata birds are invertebrates that float in the Pacific Ocean much like Jellyfish. They also enjoy small jellyfish and squid. During the migration, they eat crustaceans even though they have very sensitive jaws. When they arrive in Alaska, they collect food. When on land, the foytile does not hunt, it scavenges for plants and berries. They use their keen sense of smell to sniff out their favorite types of food.

Along with their excellent sense of smell, the foytile has many other features that make them amazing. From beak to tail they measure 10 feet in length and weigh an average of 2,019 lbs. Their thick coat allows them to be in temperature of -50 C before needing to increase their body heat. In the winter season their coat is stark white and then turns brown and gray in the spring and summer. Their tail is waterproof to keep their balance in the water. They also have tiny little paws for extra support and long arctic fox ears. They also have a lifespan that can be as long as 40-45 years, during which they will spend 37-42 years with the same mate.

The Foytile's mating season occurs in March to April, depending on the year. Their pregnancy lasts 50-60 days, during nesting season they lay roughly 100 eggs. Foytiles, being very social animals, live in "families" of 1 male and 2 females. The male and they 1st female are a pair and they defend the pups and nest when they are weak. The 2nd female helps care for and raise the pups. They are extremely dependent on these "families" to help them survive.

Overall, Foytiles are amazing animals. Their peaceful appearance and fierce instinct make them unique and worth saving. These creatures are vulnerable because of pollution and plastic in the ocean. They are in danger of becoming extinct in the next 20 years if nothing changes. So to help them, please be aware of the impact you could make on the foytile species.

The Ocelot
Zoë Sonnenberg

I was sprinting across one of Argentina's dense forests. The rabbit was fast, but I was faster. I finally caught the rabbit. It was delicious, even if my kittens got more than I did. I know that sounds mean and you're all like, "Awww. Poor bunny!" But, I am a carnivore! Anyway, back to the story. The bold black spots and stripes on my yellow fur made it look bright. I can't remember the last time I had another meal. I usually eat fauns, birds and even fish.. I've become quite skinny, but I can't remember what my actual weight is. I could be anywhere between 18 and 27 pounds, like a normal female of my kind.

I went back to my den and spent time with my kittens. I'm going to miss them when they go off on their own. They're all almost one year old. That's when they leave me. Alynia, my daughter, said she wants to live in either Texas or Mexico. Joe, my son, wants to live here, in Argentina, so he can visit me. One time they got away from a jaguar and they didn't even need my help. I MEAN A JAGUAR FOR CRYING OUT LOUD! I worry too much.. They are almost grown up. Alynia will be ready and mature at 1 ½ and Joe, will be mature at 2 ½. At least I have time with them now.

That story happened when I was young. Im 9 now and my children are well on their own. My once yellow fur is now gray with flecks of white. I am old but proud. I saved myself many times from harpy eagles, anacondas, and many many more. Hunters come every once in a while, looking to capture me for my coat, but I am still fast. I'm rather lonely since my kind are solitary animals. I am an Ocelot and I am brave.

Tundra Swans
Hana Carmon

Tundra swans are majestic snowy white creatures who soar through the sky. They have black beaks like a trumpeter swans, but the difference is, tundra swans have yellow spots on their beaks. They have feathers that seem like they are snow. Their wings are 5.5 feet long while they are about 3.9 to 4.8 feet long. They can live to be up to 20 years old in the wild. When tundra swans are babies they have grayish white feathers that are waterproof and weather resistant.

Tundra swans are omnivores, so they eat both plants and animals. They eat submerged plants, roots, grains, corn, and mollusks. They dip their heads in the water in order to get the plants underwater. They also eat wigeon grass, sago, clasping leafs, wild celery, winter wheat seeds, and pondweeds.

Tundra swans can lay up to four eggs. They lay their eggs near the shore and when they lay their eggs they incubate their eggs for 32 days. They are very territorial when other tundra swans come near their nest. They will snap at them to tell them to back off. Before they mate, they stay with another swan for one year. Tundra swans will stay together and mate for the rest of their lives. They make their nest out of moss, sticks, and grasses.

They live in many different places such as North America, Europe, and Asia. Tundra swans mate at Bristol Bay, Alaska, Bering Sea Coast, the Arctic Ocean, Baffin island, and Quebec. They migrate to the pacific coast, arctic coast, British Columbia, the Great Salt Lake, and northern California. Tundra swans usually nest in the wet artics and are usually found near the coast. They live in shallow lakes, slow moving rivers, flooded fields, and coastal estuaries.

The tundra swan is a beautiful creature. People should look at them and admire their graceful beautiful form. Their beautiful white feathers reflect the color of the snow.

Tundra Ocelot

Hana Carmon and Zoe Sonnenberg

Hi, my name is Lily and I am a female tundra ocelot. I was forced to run away from my nest when a harpy eagle attacked it. I had to hide in the pond that was near my nest. I used my white fur to my advantage. I covered myself with pondweeds and used it as camouflage. I should have been careful since jaguars and pumas prey on the smaller ocelots like me and birds like the harpy eagles try to eat me. Even the worlds biggest snake the anaconda tries to eat me. When the harpy eagle disappeared, I uncovered myself and went to the surface. I noticed it was night time so the harpy eagle must have been here for a long time. I had to use my night vision to see where I had to go. I heard my stomach growl, so I looked around for food. It took me 30 minutes to find some food. It was a rabbit fish. I had to stay quiet since rabbit fish were speedy and could get away in a minute. I accidentally stepped on a branch and it cracked the rabbit fish must have heard it since it ran away. I guess I just lost my dinner.

The next morning, when I woke up I went to a puddle to lap up water. I looked at my reflection to see my blue eyes that were usually filled with life and excitement looked all dull and lifeless. My wings were all chewed up and looked ragged. My white fur was dirty from when I was hiding in the pond and even my spots and stripes seemed dull. My beak was dirty and it was usually shiny and looked clean, but now it was all scratched up and muddy. I was snapped out of my thoughts when I heard my stomach growl. Oh yeah, I forgot I didn't eat dinner, I thought. So I went in search for some kind of food to eat. I came across a stream filled with rabbit fish, so I dipped my head in the water and I caught one in my beak. I began to eat enjoying each bite taking in the savory goodness. After I ate, I decided to make a new home for myself since my previous home got destroyed by that stinkin' harpy eagle. I went around looking for a nice cave where I could set up my nest. When I found a perfect spot, I started collecting stuff to build my nest like moss, sticks, and grass. After an hour or two, I finished making my perfect nest. When I looked outside of my cave nest, I saw it started to get dark so I called it a day and tucked myself into my nest.

I yawned as I woke up to the bright sun. Time to go wash myself off I thought to myself. I walked to a pond that was near my cave and jumped in. When I got out, I started shaking all the water off me. I was all clean. Hmmm what should I eat today. Maybe I should look in the pond to see if I can find some plants like wigeon grass, sago, clasping leaf, pondweeds, wild celery, or winter wheat seeds or maybe I can look around the forest for rodents, fauns, birds, snakes or even rabbitfish. I stared into the pond when I saw some clasping leafs. I dipped my head in the water and used my beak to grasp the clasping leafs. I started chewing it with the mini teeth in my beak. After I finished eating, I went back to the cave and started wrapping myself in old leaves and fell asleep.

The next morning, I was rudely awoken by something.

"Go away," I said to the thing that was trying to wake me up.

"Never!" a childish voice replied.

I instantly sat up from my nest to see a gray ball of fur. "Ahhhhhh!" I screamed. "Who are you?" I yelled.

"My name is Peter," the thing said. After he said that, I knew he was a boy.

"What are you?" I asked.

"I'm a tundra ocelot," Peter said in a happy voice.

Hmmmm that's weird it's pretty rare to see baby tundra ocelot since its' not even mating season and they are so vulnerable to predators I thought.

"Hello are you there?" Peter said.

"Hmmm sorry I got distracted, wait where's your mom?" I said.

"I don't know my mommy told me to run."

"Oh ok, you can stay with me."

"Really? I can stay with you? Yay, I have a family now!"

"We'll go out everyday to look for your mom and we'll never give up." I said.

From there on Peter and I would go out everyday looking for his mom until the day we found her. Everyone was happy. Peter went back to his mom and I was taken in by Peter's mother and raised as her own.

Hedgehog
Tien Brown

Peeking out of my little hole looking for predators is me the hedgehog . I don't have very many predators, though some of my predators like the European owl will somehow find a way to get to me. I know this because I have heard from other hedgehogs that some of our bones have been found in the European owls pellets. Also the badger is one of my predators, somehow they can get to us, but I wouldn't know because obviously I am still alive. Other than that I don't have very many predators.

I may look cute and cuddly, but things don't always look the way they seem. My spikes are sharp, but underneath on my belly and on my face I don't have any spikes, so I curl up when I sleep to protect myself. I have small ears and small legs as well as a small tail. I weigh up to 14 to 39 oz. I am so small that my size is relevant to a tea cup, and sadly I have long weak claws. One last thing about my appearance is that if you see me in your garden eating my prey you will see that I have surprisingly sharp teeth.

I eat a variety of food. I like to eat insects, frogs, snails, snakes, toads, bird eggs, mushrooms, grass roots, berries, melons, and mice. One of the melons that I eat is watermelon, which is by the way delicious. My favorite out of all of these is bird eggs. I am known to be eating the pests in some of people's gardens. Just one last fact about about this topic before I move on is how I eat. When I catch my prey I use my sharp teeth to grab my meal and I eat it really fast so that it doesn't get away.

Cancer is very common for my kind. The most common is squamous cell carcinoma. Squamous cell spreads quickly from the bone to the organs in hedgehogs, unlike in humans. Surgery to remove the tumors is rare because it would result in removing too much of the bone structure. Some other diseases I can have is fatty liver diseases, which is believed to be caused by a very bad diet [better keep a good diet cause I don't want that]. My kind will eagerly eat foods that are high in fat and sugar. Having a metabolism adapted for low fat, protein rich insects, this could lead to having common problems of obesity.

All in all my kind is special in a way because I am not a porcupine even though if I might look like one. I am a unique animal who doesn't throw out my spines but keeps them until they fall off. Don't forget to say hello to me if you see me in the wild.

Indian Peafowl
Leah Phu

I am national bird of India. I am a large brightly colored bird. I am a type of peacock that originates in India. My head, neck, and breast are a glossy, iridescent blue (which means when seen by different angles it changes colors), with white patches above and below my eyes, along with a crest of upright, blue feathers on the crown on the top of my head. In contrast, my back and wings are greyish-brown with brown barring. Undoubtedly, the most striking feature of my species is the long train of feathers at my rear end, which, I can encompass nearly two-thirds of my total body length.

You can find me in India, Pakistan, Sri Lanka, southeast Asia, and central Africa. My kind is often found in open areas like parks and trees that are close to water sources and also live in deciduous tropical forests. Because we are not threatened by the presence of humans, we are often found to live near farms and villages. I also can be found in moist and dry forests, but I can adapt to live in cultivated regions and around areas populated by humans. In many parts of northern India, I am protected by religious practices and will forage around villages and towns for scraps.

I am omnivorous, which means I eat many things such as insects, plants, berries, small amphibians, reptiles, and mammals. The main part of my diet is fallen berries. I forage for these berries and other foods in the early morning and shortly before sunset. I retreat to the shade and security of the forest for the hottest part of the day.

I run more than I fly. The only time I fly is when I have to cross a river or ravine, when trying to escape predators, and to roost up in trees. I warn other Indian Peafowls when danger approaches with loud, shrieking cries and honks. I stay in small flocks which are called harems, of 1 male peacock and 3-5 female peahens. The most common predators I have are tigers, leopards, civets, wild canines, and mongooses. My train of feathers is extremely problematic because it sometimes blocks me from being able to see dangerous predators that might be coming up from behind. When predators try to tug on my train, I often get lucky; sometimes my feathers simply drop to the ground enabling me to bolt from the scene immediately. In some areas where I am not protected by law, humans hunt my kind to make "peacock oil".

Not only am I India's National bird, I am highly regarded in the Indian culture. I am often depicted in temple art, mythology, poetry, folk music and Indian traditions. Not only am I a significant animal in Indian culture, I also play an important role in Buddhist philosophy and Greek mythology. In the past ,I was often kept in menageries (menageries means a collection of wild animals kept in captivity for exhibition) and as ornaments in large gardens and estates. Even today I am still used as a symbol of beauty in any culture.

Indian Hedgefowl
Tien Brown and Leah Phu

I am a the Indian Hedgefowl...

I am a strange kind of mammal species. I am common in North America and Asia. You can find me in open areas and underground. I have a very strange time being with other animals because they are all alike and I am the odd one out. My appearance is not what a person would expect of me. I am not like other animals.

I have a strange appearance to some people. I have a Hedgehog like nose and whiskers. My eyes are small and glossy and I have a small hedgehog shaped body. My tiny body allows me to live in small places and for me to hide in holes to get away from my predators. I also have feathers that have spikes between them to defend myself from my predators. My beautiful feathers are similar to a peacock's.

My predators are European owls, tigers, leopards, civets, wild canines, and mongooses. These predators are very hard to escape from, because my bright colored feathers make me stand out. Where I live is an advantage because of all of the trees and small places to hide. I can hide in the bushes and climb the trees to get away from my predators. Also, when predators are nearby I fluff up my feathers and spikes like a pufferfish to scare them away.

I love to live in both moist and dry areas. When living in moist climates, I live in grassy fields and by freshwater. When the weather is dry, I try to find a good resource of water and food. Even though I might live in moist and dry spots, I still have trouble finding food at times because we travel at slow speeds.
I eat many foods like plants which include berries, melons, and more. I also eat small mammals, reptiles, and insects when I can catch them. You might even find me strolling in open parks.

Even though I am not a very common animal I am still a very special one. My features mostly tell about me. I have spikes like a hedgehog and feathers like a peacock. All my features together make me a one of a kind animal.

I am the Indian Hedgefowl...

The Indian Rhinoceros
Jon Lao

In an area in the Indian subcontinent lies a small baby rhino. But this rhino isn't an ordinary rhino, it is the Indian Rhinoceros, also known as the Great horned Rhinoceros. When the baby rhino grows up, it will feed on grasses, trees, and other fruits and plants because Indian Rhinoceroses are herbivores.

Apparently, the baby rhino never knew that its life was endangered because of poachers who hunt for rhinoceros horns. Even officers in India shot more than 200 of the rhinoceros species. Since then, rhinoceroses live and travel around floodplains, regions with tall grasses, adjacent swamps and forests, and areas in Africa and Asia.

The baby rhinoceros will move to different areas with its father and mother. Sometimes rhinoceroses are found in big groups. Those big groups are called crashes. The baby rhino and his family will eventually move into a crash and stay with that group for a while.

When it comes to the history of the rhinoceros species, people talk about the Indian rhinoceroses' great stories in the past. Indian Rhinoceroses numbered fewer than 200 animals in the early 1900s. Some biologists believe that an Indian Rhinoceros can live up to 35 years. If it's in a strict secured environment, it may live up to 40 years. So the history of the Indian Rhinoceros species is a great topic to put in conversation.

Have you noticed the rhinos armour-plated appearance? The Indian Rhinoceros has large folds of skin at the joints and thick neck rolls at its neck that combine with the knob-like lumps and large plates that cover its gray body to give the rhino an armour-plated appearance. As this gray small horned creature matures, it will grow to have this armour-plated appearance.

It's interesting that Indian Rhinoceroses aren't like other rhino species that use their horns for defense. Instead, Indian Rhinoceroses use these long canine teeth in their lower jaw to protect their young and defeat their foes. In rare cases, when tigers, hyenas, and lions come to find a rhinoceros snack, they'll use their horns for defense. They probably only encounter these creatures when predators seek to steal babies.
Depending on their habitat Rhinoceroses can be solitary creatures. As solitary creatures, both male and female rhinoceroses mark territory. Males defend their territory while females nurse their young. Sometimes, rhinos congregate in bathing areas and live near zebras and horses.

The baby rhinoceros grew up and left the crash. Using his survival instincts, we can be sure that this Indian rhinoceros led fabulous life. Later in his life, he did live a successful life! He was able to mate with another Indian rhino and had children to take care of. Humans knew that the rhinoceros species was endangered. So some rangers found this family and kept them in a safe enclosure. The Indian rhino lived until he was 40 years old. This was the age that the scientists estimated a rhino could live in captivity. No one ever knew his name so they named him Rico, Rico the Indian Rhinoceros.

That my friend, is the life of an Indian Rhinoceros. Indian Rhinos are still endangered today, so do your best to prevent the Indian Rhinoceros population to become even more depleted.

Lionfish
Nicholas Flynn

    Beyond what the eye can see there is an underwater world, and if you look hard enough you can see many of the beautiful creatures that live within the coral reef...

    I am a Lionfish. There are all different shapes, colors and sizes of me. My body can range from 5 centimeters all the way to 45 centimeters in length. I weigh 1.5 pounds but I can weigh anywhere between .88 ounces and 2.9 pounds. I am brown and maroon with white stripes, but other lionfish could have stripes that are such a dark brown that it sometimes looks black. Some can even be red with white stripes. I have large number of venomous spines and fins that look like tethers waving through the water as I swim. My mouth is gigantic in comparison to my body. This usually allows me to swallow other small fish whole. My body can withstand many different temperatures, depths and saltiness of water, so I have the ability to live in many places.

    I can be found in a huge variety of areas around the world. I can usually be found in warm tropical waters near islands. In 1992 my aquarium in Florida was destroyed by a massive hurricane and many fish, including myself, were set free into the Atlantic Ocean. We eventually spread throughout the Caribbean and all the way up the east coast of the United States as far as North Carolina. But this isn't the only place that lionfish inhabit. A lot of lionfish can be found anywhere from western Australia and east Malaysia to French Polynesia and the United Kingdom's Pitcairn Islands. Some lionfish can even be found around Japan and southern Korea and all the way back south to Lord Howe Island off the east coast of Australia. Within these areas around the world lionfish can be found in water depths anywhere from 1 foot to 1,000 feet. I prefer to live on the coral reef, but many other lionfish live on the rocky areas on the ocean floor, mangroves, seagrass, or even artificial reefs like shipwrecks.

    As a poisonous fish, I don't have many predators. The predators I do have, have a difficult time seeing me because my colors blend into the reefs where I hide. My poisonous spikes have a paralyzing toxin which I use when pursued by a predator, but I rarely use this poison when hunting for prey. My predators include Caribbean sharks, groupers and Moray eels. Sometimes I have even been know to eat my own kind. Another predator of mine is Humans. Humans often capture my kind as interesting looking pets and as a delicacy to eat. In the Caribbean, predators are having a difficult time hunting my species because we are so new to the area and they haven't adapted to our tactics. Although my species does have predators, we reproduce so rapidly that our numbers aren't threatened.

    I am an adept hunter. I use my fins and huge mouth to trap and swallow small fish whole. I usually like to eat juvenile fish, shrimp and other small crustaceans. Not only are other lionfish a threat to me, they are my prey as well. I'm not the friendliest fish in the sea, I typically spend my day hiding in reef crevices and tunnels, only coming out to feed at night in deeper waters. The juvenile fish that I prey on never had the opportunity to reproduce and this ends up eventually depleting that fish's population. Many of the fish that I eat play an important role in maintaining reef health. For example, I love to eat parrotfish, and parrotfish help to clean the reef. If too many parrotfish get eaten, algae can cover and smother the coral. By feeding on native fish populations, I end up disturbing the balance and the health of the reef may be seriously affected. Even though I am known to humans for my beauty I am a huge threat to my own habitat and the other sea creatures that also call it home.

The Indian Lionoceros

Jon Lao and Nicholas Flynn

If we look into the Maldives, we will see the Indian Lionoceros. This rare creature is almost impossible to find in this world. People have only seen around 4 to 5 times in the wild. Thats how rare this animal is. What you don't know is that its actually a mammal even though it swims like a fish.

When baby Lionoceroses grow up, they do not have spiny armor plates just yet. instead, they are just plain gray with stripes. An adult Lionoceros has spines on its armor plates that are flat when not in use so it can surprise their foe with a pointy and poisonous attack. But it only uses this special move when its defending itself from hyenas, Lions, Tigers, and other foes. When its charging at its foe, it will use its Horn that has a small spine from its armor attached to it so that it can paralyze and poison a foe. It will keep its spines flat like flaps so it can not be noticed by predators and prey. Its armor is also maroon with thin gray stripes. This is unique because it helps the lino catch prey by camouflaging with its surroundings.

The lino eats grasses, bushes, tree leaves, and sea grass if necessary. Sometimes it even preys on small fish when its swimming. The lino usually scavenges for food in groups of 4 to 5 called crashes.

Lionoceroses are very social creatures. Not only do they search for food in their crashes, they also travel in crashes and they defend each other from predators together. If Lionoceroses are in crashes, they can probably live up to 50 years. When alone in the wild, they can live up to 45 years.

Lionoceroses migrate throughout India. They transfer to the Maldives during the summer so that they can relax. The Lionoceros lives near oceans and grasses in areas in India. When it travels, it will swim in the water to avoid dangerous predators on land. When it swims, it will kind of hide its legs under its body so that it can swim swiftly and smoothly. this is very unique feature because most land dwelling animals rarely swim in order to migrate.

The Indian Lionoceros will take shelter in a nearby hole or cave when a storm appears. if there are no caves nearby, it will run as fast as it can or it will try to bury itself underground. It will then migrate to a different area in India so that it can avoid the storm. This skill is very useful to this animal because it can take shelter in different ways.

Not only is the Indian lionoceros a rare creature, it is also an extremely unique animal with different skills and adaptations. Because of its rarity, people all over the world are in search of this amazing species.

Axolotls
Grant Edwards

Deep in the forests of central Mexico in the lake complex of Xochimilco, there was an axolotl named Bobet. She was best friends with another Axolotl named Bob. Bobet was a bright shade of pink and Bob was a dark shade of blue. They each had four legs and a big tadpole tail, which they have had since they were young. Bob and Bobet each have a set of leathery gills that protrude from their heads rather than having slits on their necks. Their 18 year life is plenty of time for them to explore the lake bottom. These two axolotl friends usually had a basic routine of looking and smelling for small pieces of fish, crustaceans or insect larvae, but today was different.

Bob and Bobet were walking on the bottom of the lake that they lived in when a worm caught Bobet's eye. Before Bob knew what was happening Bobet darted off and chased after the worm. It got away before she could catch it. Bobet had followed the worm all the way to the surface of the water and next to the shore. The worm appeared once again and then disappeared into the rocks alongside the lake. Bobet hungrily chased after it. Legend has it that if an Axolotl goes onto land for long enough, it will turn into a salamander. The difference between an axolotl and a salamander is that a salamander lives mainly on land and cannot be in the water for extended periods of time. Bob wasscared for Bobet. He didn't want her to turn into a salamander and he didn't want her to get snatched up by a bird. Bob didn't want to lose his best friend. Bob waited for Bobet in the water by the shore all afternoon, with still no sign of Bobet.

The next day Bob walked along the bottom of the lake to get his mind off of Bobet. As Bob was walking, there was a piece of plastic that Bob had mistaken for a crustacean. When he tried to eat it he started to choke and spit it back out. Bob thought to himself that he had been seeing more and more plastic and other trash around the lake. Bob didn't know of anywhere else in the world that Axolotls lived, so the pollution that humans were leaving behind could become a big problem. When he was walking by yet another discarded piece of glass Bob rubbed against it and his leg got cut off. Luckily, Bob has the ability to regrow his limbs, so he wasn't worried.

Meanwhile on the shore of the lake complex Xochimilco...
Bobet has turned into a salamander and now can't go back the water to see Bob. Bob was at the very edge of the water hoping to see Bobet but all he saw was his reflection. Bobet was also looking at the water looking for Bob. Bob stuck his little head out and realized it wasn't his reflection, it was Bobet! She didn't look much different, but she still couldn't live underwater the way that Bob could. They met up almost every day from then on.

LoggerHead Sea Turtle
By Jake Richter

    I am a loggerhead sea turtle. I live in coastal waters and search for food up and down the coast of Baja California. I have the hardest back shell in the world. As an adult, I weigh 80 to 200 kg on average. I also swim up to 15 miles per hour.

    I am very adaptable because I can live in both deep and shallow waters. I live in shallow waters in coastal places because coral reefs. I also can live very far out at sea even hundreds of yards out! I live in the Atlantic ocean near Mexico. Every April I migrate to either Florida or North Carolina depending on where I am. I migrate to either one of these places to lay eggs or to find warmer waters during colder seasons. I migrate anywhere from 100 to 1,000 miles. Mostly, only the female do this because they lay the eggs, but sometimes I tag along.

    I like to eat many things. Those things include shellfish. I like shellfish because I can crack hard surfaces with my strong beak and teeth. Don't get me wrong though, I also love soft foods like jelly fish, star fish, moss, seagrass, and sponges. I also like to eat sea cucumber. I have to be careful though, because some animals are poisonous.

    Life is all good and everything but, it is still very dangerous. I have to watch for sharks, seals and human fishing nets. But even on land, I have predators. There are red foxes that can attack while we nest and when the babies are making their ways out to sea for the first time too. Seagulls also try to get the babies when they are going out to sea. To protect myself, I swim away from my predators. This is unusual for turtles because most of them go into their shells to protect themselves.

    These are just some things about loggerhead sea turtles like where we migrate, what we eat, and our predators. We are related to other sea turtles but we are different in many other ways.

Turdleotoltl
Grant Edwards and Jake Richter

The abnormal creature you see on the opposite page is an amphibian, which means it can go on land and in water. It can go in fresh and saltwater as well. It is known as the turdleolotl (tur-da-low-dull).

The turdleolotl has very complex features that it uses to survive. One of these features is the little gills that protrude from their heads. These small gills help them stay underwater for very long periods of time. The detailed pattern on their shell helps them camouflage into any surrounding. Their skin also varies in color depending on its environment, however, its shell stays the same. It is always green. From head to tail, the turdleolotl is 5 feet long and weighs 60 pounds. Despite it's larger stature, the turdleolotl is an agile swimmer. It can swim majestically at fast speeds because of its long paddle like fins. This amazing amphibian has the ability to grow back one of these limbs if it loses it. If this does happen, they can only regenerate their body part one time. The turdleolotl also has a beak to help it catch and eat its prey.

To catch its prey, it dives down at a fast speed and snachtes it with its beak. Once it has the prey in its mouth it crushes it. Some of the prey it catches includes, crustaceans like crabs and lobsters.

It lives in the lakes, streams, and in the Gulf of Mexico. It is interesting that it can only be found in one country in the entire world. At night, turdleolotls sleep in the water or they sleep in little holes they dig on the shore. They mostly alone, but are sometimes found in groups. To survive, it does a lot of things to protect itself.

The turdleolotl can go in its shell to protect itself from its predators. Some of its predators include great white sharks and sea lions. In addition to natural predators, humans are also a big threat to this species. Boats and motor oil from humans are one example of how people harm their habitat. They are actually on the endangered species list because their habitat is being destroyed by these types of actions.

The turdleolotl is a very rare and endangered creature. It lives in only one place in the entire world and this place is beginning to become over-populated and over-polluted by humans. Turdleolotls are in danger of becoming extinct if nothing changes.

Sea Star
Carina Chan

Sea stars are often thought of as just sea creatures shaped like stars, but underneath that simple star shape is the life of an advanced animal trying to survive. They have many unique adaptations to help them survive and thrive in the kelp beds of the ocean.

A common sea star can grow up from 4.7-9.4 inches or 12-24 centimeters. They can weigh up to 11lbs or 5kg. Sea stars can also live to be 35 years old. Sea stars actually don't have blood, their "blood" is filtered sea water. At the bottom of a sea star it has a round center and right in the middle is where its mouth is. The arms grow out of its round center likes spikes on a wheel. To help itself eat it has suction cups on its feet to pry open clams or oysters. It is a unique colorful animal. Sea stars have colorful skin that helps them camouflage or intimidate their predators. Another way it can defend itself is by using their spiky skin. Its skin can also help it smell food.

Sea stars are carnivores so they eat meat. Some of the things they like are; mussels, clams, oysters, shellfish, and sometimes other sea stars or any small animal. They don't just eat meat they also eat coral, sponges, and sea anemones. Sea stars is that they can consume prey outside of their bodies.

Sea stars need to keep themselves alive with their amazing survival skills. If a sea star's leg gets eaten or injured by a predator it will grow back. They have spikes around their mouth to keep their insides safe. Its firm, bony skin keeps itself protected from most predators. Though a sea star has many advantages to its survival life, it travels very slow, it travels less than a foot per minute. Sea stars are currently not endangered, but they can be affected by pollution and temperature change. Some of their predators are manta rays, sharks, large bony fish, and sometimes larger sea stars.

Sea stars live in the Pacific coast of North America near kelp beds, sea grass, coral reefs, and tidal pools. Though they are most often found near rocky shores. They have a wide range of where they live, they are found from tropical homes to the cold sea floors.

Sea star have many cool qualities and advantages in life. Although it doesn't have the typical defense mechanisms as other sea life, it does have its own unique ways to adapt and survive. It eats many interesting things and can be found in many different places. Sea stars are not boring and lifeless like any plain object, they are shining stars but instead of being in the sky they are in the ocean.

Chinchilla
Angeni Nettles

Chinchillas are mammals that come in many different colors. Some colors are gray, black, white, brown, blue, and even violet. The chinchilla has two very big ears and two small eyes. Chinchillas are cute, fluffy and amazing creatures that look similar to but are different than mice.

They are close family members to the mouse, but you can tell them apart by their tails. Chinchillas unlike mice who have small thin tails, have fluffy tails. Their big fluffy tails, stretch up to 5-6 inches long. Chinchilla are small rodents that are 12-13 inches long the size of a ruler.

These small creatures, live in harsh, cold, and windy climates that provide them a food source and shelter. They live in the Andes Mountains and have soft thick fur that helps them stay warm and dry. In the mountains, cardonplate is a type of of cactus that gives them m ore food and water as well as a home to live in. In addition to the cardonplate, the chinchilla will sometimes make burrows in the cracks that are made in the earth. In this habitat they can find food such as, scrubs, berries, seeds, cacti, roots, and insects. They don't get very much water so the rely on the dew in the morning to stay hydrated.

These cute little rodents are very social creatures who like to live with others. They stay in herds that can grow up to 15-100 chinchillas that are sometimes called colonies. They interact with one another while grooming, mating, and communicating. Chinchillas are very curious animals that will jump up to 6 feet so they can see their surroundings.
They have to use their adaptations to protect themselves from predators. The main predator of the chinchilla is the fox. Other predators that will feed on chinchillas are owls, cougars, mountain lions, and snakes. Since chinchillas have very loose fur, they are able to shed their fur when grasped by a predator. They will also jump up very high to get away from predators near by.

Even though chinchillas are small creatures they have big personalities and aspects that help them survive in their environments. Even though you may think a chinchilla is a mouse at first, sneak a closer peek but not too close so they don't jump away.

Starchilla

Carina Chan and Angeni Nettles

Starchillas are in danger of becoming extinct because their fur changes colors based on their mood. When they are killed, their fur stays the color it currently is forever, so a fur coat made from starchilla fur will never lose its color. The starchilla is an amazing creature and should not be hunted for its fur.

The starchilla has many appearances and features to help it survive in the wilderness. It has the head, body, and arms of a chinchilla, but also the arms of a sea star. It comes in many colors including; gray, black, brown, white, blue, and even violet. When it goes into the water to cool off, it can change its colors to bright or dull to adapt to different surroundings. This camouflage adaptation helps it blend into its habitats and helps it get food. Its long fluffy tail is about 5-6 inches in length. Its size is 4.7-9.4 inches long, and it weighs 11 lbs or 5kg.

Starchillas feed on the cordon plat, a type of cactus. It can also provide them food and a shady place to live. They will also feed on clams, insects, berries, and scrubs. To hunt for living prey, it first hides and then only when the prey gets close, it pounces on it! When they eat an insect they eat the whole insect because it's very accessible, and gives them a lot of protein. The starchilla uses its suction cups on its hands to pry open a clam's shell then suck out the insides. They'll scavenge for scrubs and berries and put them in their home until they have enough.

If they can't find a proper place to live, they will make burrows in the cracks in the ground. Starchillas will usually stay near places with cacti and water. They live in deserts in the southwest part of North America. They migrate in the winter to South America where the Andes Mountains are because the weather is warmer and they can mate and breed by spring time. Starchillas like to migrate in big groups because they are very social. It needs to eat a lot of food before the migration season, because it travels slowly in water.

It might travel slow in water, but it can reach high speeds on land. Whenever it eats clams or needs to get some berries up high, it uses the suction cups on its hands. The starchilla's ability to change colors helps It camouflage or strike predators. Its bony skin on its legs help protect the starchilla's legs from most predators. When the starchilla is grasped by a predator, it is able to shed its skin very quickly. The predators of the starchilla are; large sea creatures, foxes, snakes, owls, cougars, and mountain lions. It'll jump 6 feet in the air when it sees or hears nearby predators. It can also jump to see its surroundings. This is only one behavior of the starchilla.

It is also a very social creature that likes to stay in herds, the herd can have up to 15-100 starchillas in it. A starchilla can live up to 60 years. Since starchillas are so social they like to interact with each other while grooming or mating.
Starchillas are rare and interesting mammals. They are endangered and getting closer and closer to extinction. Since starchillas are magnificent creatures they are worth saving.

BULL SHARK
Nolan Edge

The bull shark is everywhere you wouldn't expect them to be. Bull sharks can migrate very far and they can swim in saltwater and freshwater. Some bull sharks have been found the freshwater Potomac river, the Mississippi river and even in Lake Pontchartrain. Most bull sharks can be found in warm fresh water and warm ocean water. Some bull sharks live in rivers and prowls into shallow waters looking for food. Some bull sharks migrate 2,300 miles from the upper amazon river to the sea.

After about ten years bull sharks reach maturity. Adults are usually around 3.5 meters (11 feet) long and weigh about 300 kilograms (660 pounds). Most of the males live for 13 years and females live up to 17 years of age. The bull shark's classification is a fish because they have gills and they live under water. They are grey on top and have a white underbelly to help with camouflage. They have three rows of extremely sharp teeth.

Bull sharks have a huge appetite and eat almost any creature they can find. They mostly eat boney fish, small sharks, mullet fish, tarpons, and catfish. They also could eat menhaden, gar, snook, jacks, mackerel, snappers, and school fish. When a bull shark hits its prey it twists and pulls to get bigger chunks of flesh off. Some sharks hunt with others but they mostly hunt alone. Sometimes they hunt during the day but most of the time they hunt at night. Bull sharks are very territorial.

Bull sharks do not have many predators. The only threat they have is humans, mainly fishermen. Fishermen catch bull sharks and sell them as food. They are not only hunted for their meat, they are also hunted for their hide and oil.

Although bull sharks are hunted by humans, bull sharks are actually the most dangerous shark in the world for humans. The reason they are so dangerous is because they like to live in shallow waters and they are very territorial. Bull sharks usually attack Humans that are recreationally swimming in shallow waters because they are defending their territory. Humans should be cautious of finding a bull shark where they are least expected.

New Guinea Singing Dog

Libby Williams

New guinea singing dogs are the most domesticated of wild dogs, but their population is actually quite small. There are only 100 new guinea singing dogs left in captivity and a small unknown amount in the wild. People have begun breeding them to restore this animal's population and now many people own them as pets.

New guinea singing dogs might look like your average household dog, but really, they are wild dogs. The New guinea singing dogs has a doubly thick coat and a bushy tail. Their coats can be found in a red/brown or a brown/tan, but they are most commonly found in a tan color with white markings on the tip of the tail, feet, chest, and underside of the chin. They have wide cheekbones, narrow muzzles, and tulip shaped ears that stick up. They average 17 inches at the shoulder and weigh 25 pounds. New guinea singing dogs' spines are very flexible, giving them the gift to climb up trees in order to stalk their prey.

New guinea singing dogs live in the mountains and thick forests in Papua New Guinea, which is located on the continent of Australia. They have survived in their natural habitat for years, but now their population is dying off. Nobody knows why they are dying off. They are very rare, so people have been taking them into captivity or their homes to breed them.

New guinea singing dogs eat small birds, fruits, small reptiles, and small mammals. Some of their favorite things to eat include yams, sweet potatoes, small snakes, rabbits and mice. Due to being wild animals, they have to eat what is in their habitat. To hunt, they climb up trees and can watch their prey from above. They get their water from near by water sources like the Gulf of Papua New Guinea, and Torres Strait, which is between Papua New Guinea and Australia.

The new guinea singing dogs hunts alone. They are solitary animals, and only stay with mates to defend their territory. They live with their mates and their pups (babies). The male and female both participate in raising the pups. They used to be companions for a tribe in New Guinea but now they are just solitary.

Due to taking the animal from its natural habitat, it is believed to be extinct in the wild now or very few left in the wild, but organizations are trying to restore this magnificent breed of animal by breeding them at home.

New Guinea Bull Shog

Libby Williams and Nolan Edge

All mythical animals can live in the water or on land, but some can go live in both environments. The New Guinea Bull Shog is one of these animals. This animal is a strong hunter, unique looking, and can often be found alone. It was created by combining the New Guinea Singing Dog and the bull shark.

The New Guinea Bull Shog mostly eats small animals and fish. Mullats, mullet fish and a rat, catnakes, catfish and a snake, and yamirds, a yam fruit and a bird, are some examples of what they eat. They eat pretty much anything that is in their way, that is not too big. Their spines are very flexible so they can turn quickly to catch their prey. Their prey can mostly be found in New Guinea.

They live in the warmest parts of the forest and mountains in Papua New Guinea, and Australia. They swim in freshwater lakes and saltwater lakes located in the mountains. They make dens out of sticks and brush to blend into their environment. They can also make a big hole to live in and the roof that is made out of parts of trees and leaves. They dig a hole by the water and make a den, then they dig another hole to the lake so they can swim in the lake. They stay in their dens when it is very cold, but still can go outside. The New Guinea Bull Shog can go in a lot of lakes and they have legs so they can go on the land.

New Guinea Bull Shogs are very territorial over land and food. They live alone, unless with a mate. The female makes the houses and the male goes to hunt. They spend most of their time taking a swim and protecting their land. They hunt alone because they are greedy. If they were to hunt with another New Guinea Bull Shog, they would fight about who would get the food because they would both want it for themselves. The New Guinea Bull Shog uses its features to catch its food.

The Shog's have wide cheekbones, narrow muzzles, and tulip shaped ears that stick up. They have white markings on the tip of their tails, feet, fins, and on the underside of their chins. Their spines are very flexible so they can turn quickly to catch their prey. The New Guinea Bull Shog has three layers of very sharp teeth. They weigh around 325 lbs. Their body is slick and slimy, but their head and neck are furry.

This is how the New Guinea Bull Shog survives and lives. It shows their lifestyle and what they do to hunt, their appearance, and their size. The New Guinea Bull Shog is a talented animal at hunting and at survival.

The Platypus

Cameron Smith

The platypus is an odd animal. They have the feet of an otter, a tail of a beaver, and the bill of a duck. Platypuses are not secret agents from T.V., but they are so venomous they can paralyze a grown man. These five pound creatures have many predators that they have to face. Their predators include hawks, snakes, water rats, dingoes and eagles.

The platypus has many skills for survival. It has a small barb on its foot that only males have. It can kill a small dog with the amount of poison that it injects into their body. The platypus has webbed feet used to swim, but when they are on land they use their claws to dig holes and to run. They have a tail used for steering in the water and to store their fat. Platypuses have more than one receptor in their beaks used for locating muscle fibers in their prey's body. They use this to locate their food. The platypus has two layers of fur, one for warmth and one for keeping the water out while they are swimming.

The platypus is a very hungry animal. It can eat its whole body weight in under 24 hours! The platypus' diet includes, insects, larvae, shellfish, clams, worms, and crayfish. They eat their prey by grinding them between plates inside of their mouth. The platypus finds its food by sweeping its beak through the water rapidly to pick up a signal in its beak.

Platypuses live in Australia, near swamps. They dig holes for their young when they are just born. They lock the babies in the holes until they are strong enough to survive on their own. They dig holes with one entrance underwater and one entrance above water. They live alone after several years with their moms. Platypuses lay eggs and they produce milk even though they are semi aquatic mammal. They hunt in the night and they burrow for fun.

As you can see, the platypus may look like a weird animal but it is actually a very intelligent and cool animal that is capable of doing a lot. Sadly, they are slowly becoming extinct because of human pollution.

ATLANTIC PUFFIN
Nathaniel Houck

When you look at an atlantic puffin, you would probably think, Hey, look at that strange penguin! but you're wrong. An atlantic puffin is a bird that can fly, but spends most of its time underwater. It is a bird that looks like a penguin with a clown mask on. Its beak is a beautiful orange color but when it first hatches, it is a dark grey color. There is a mix of black and white with an orange bit of feathers on its face. Its eyes are always the same red-orange color with large black pupils in the middle. There are also a black set of feathers under the eyes.

The atlantic puffin lives in cold places near the Atlantic Ocean. The nests are made in cliff faces so predators can't get to them. The puffin then leaves its nest and migrates to places with warmer temperatures and comes back to the nesting site. It will often will use the same protected nest it used with its mate the year before.
Although these nests are protected from predators, that doesn't mean that puffins are completely safe. Some of its predators are, gulls, cats, dog, snowy owl, arctic foxes, bald eagles, and foxes. Puffins are also very vulnerable to the pollution. Nets, bottles, and plastic could choke and entangle puffins.

Puffins dive underwater to find food such as herrings, crustaceans, and sand eels. It hunts by diving underwater and catching its prey in its mouth. Male puffins usually do all of the hunting. They often will fly up to 100 km away from home to hunt. After they hunt, they fly back to the nest they made with their mates. They then divide up the food between the mates evenly.

So now that you know about the atlantic puffin, I hope you remember that a puffin is not a penguin, but in fact, a bird that flies and swims in the Atlantic Ocean. Puffins are social seabirds that look a lot like penguins, but are in fact their own unique species.

The Atlantic Puffapus

Nathaniel Houck and Cameron Smith

The atlantic puffapus is a very strange animal. It is a mixture between a puffin and a platypus. It is definitely something you would not see in your ordinary day, but if you do see one here are some facts that you should know.

The atlantic puffapus has a very hearty diet. Its diet includes crustaceans, small fish, herring, clams, mice, chinchillas, and squirrels. They hunt by diving underwater and grabbing the fish that they eat with their talons. When hunting land animals, they fly up really high, and when they spot their prey they swoop down and grab it by the neck using its talons, choking it, and making it so it is dead by the time it gets back to its nest. But, it also has predators it needs to face.

In order for it to survive it needs defense mechanisms. It uses its retractable claws and its venom to ward of its predators. Its predators include cats, large birds, and foxes. When it feels threatened, it will scratch you with its claws. If you don't go away, it will attack you with its hidden barbs on its feet. This special barb is filled with poison. It has so much poison it can paralyze a human and can kill a small dog also known as its predator the arctic fox. It also makes a noise calling for reinforcements when in trouble. It does not have many predators due to its remote habitat.

These odd animals live in the tundra up north near Alaska. They live near water and they make their homes in the sides of mountains. Their home has only one entrance with a floor covered in shrubs. This helps make the babies feel comfortable. The atlantic puffapus lays up to two eggs each year. They put their babies in the hole for up to two months until they are ready. They camouflage their children by using sticks, snow, and rocks. It uses its claws and beak to work around most materials.

The puffapus has a lot of good ways to adapt to its harsh and cold surroundings. It uses its claws in situations besides just protecting itself from predators. It can also use it for digging in its habitat. In its habitat, it can find food it needs to survive. It has a beak for grinding its food before it eats it. They have plates in their beak used for munching on its food. It has these because it doesn't have teeth. It uses its wings to fly, which has a span of 3 feet. Its feathers help glide through the air and the water with grace. It can fly up to 50 mph at its fastest. It uses its tail for steering underwater. It can hold its breath up to 4 minutes. This is useful because it goes fairly deep into the sunlit zone. It goes deep due to its food being deeper down in the ocean. Along with its defense it is a very beautiful creature. The atlantic puffapus has an orange beak. The beak is a mixture between a platypus's beak and a puffin's. They have brown webbed feet used for swimming. They also have a long platypus tail. They have black feathers with a hint of brown color. On its back their is a set of black wings. If they are the leader of their pack, they have a brown spot on their forehead. Adults can weigh up to 7 pounds. Underneath their layer of feathers there is a thin layer of black fur, this helps with keeping warm.

All in all, the atlantic puffapus is a strong and amazing creature, and are not well known around the world so if you see one spread the word and remember this story.

African Elephants
Michael Tucceri

Imagine an elephant as heavy as seven cars. Well, that's how heavy an adult African elephant weighs. Elephants may not look like it but they are actually gentle and easygoing creatures. They are one of the most amazing animals you can find today.

The African elephant is the biggest land animal living today. It can weigh up to 14,000 pounds. The trunk of an adult African elephant is about seven feet long (2.1 meters). They have trunks to eat and slurp up food and water. The ear of an African elephant is about four feet long (1.2 meters). When an elephant is feeling shy or submissive, its ears will move down. They also have tusks that are grey, which sometimes they use in fights. These amazing animals can be found in certain parts of the world.

Elephants can be found on many different continents. African elephants live in Africa and on savannas. They can also be found in India and China. They like to live in swamps, forested areas, and deserts. Unfortunately, elephants are going closer and closer to farms because they are running out of land and need more food.

Elephants eat a variety of food. They also eat grass, flowers, bananas, and fruit. In addition, elephants eat shrubs and bamboo. Elephants enjoy the taste of juicy tree sap, bark, roots, stems, vines and, shrubs. They also eat salt and can tell when they've had too much.

Elephants are very intelligent animals. They can be trained to pick up objects. They have a great memory because their brain is really big. They also can be trained to do stuff for example they can pick up a stick and scratch their back or use it as a fly swatter. Elephants have also picked up logs and used them to dig. They have been trained to clean up stuff and to put things in a pile. They can be trained to paint as well.

African elephants can do some amazing things that many other animals would never be able to do. They are truly gentle giants that can be extremely productive and smart creatures.

Meerkats
Jonas Bagaoisan

Some humans are a part of mobs and this is something they have in common with meerkats. Meerkats are one of the only wild animals to live in groups, called mobs, sometimes as large as 40 meerkats. They chose to live in these mobs to help and protect each other. A meerkat might be mistaken for another animal because of how it looks.

The meerkat has many similar features to a yellow mongoose, but they are actually very different animals. The different features between a meerkat and a yellow mongoose are that a meerkat is skinny and has tan fur. They also have black stripes on their backs. Another feature is, a meerkat's height is 12 inches long like a ruler. When they're standing up, they are 30 cm. long. Meerkats are are also known to help each other out when a predators is near by.

The meerkat lives in really hot and sunny places in Africa. They can be found in many parts of Africa. They will be found in South Africa, Botswana, Mozambique, and Zimbabwe. When they leave their habitat after three years, other animals take the habitat and live there, but they only live in plains. When a meerkat shares its habitat they mostly share it with the yellow mongoose. Meerkats also need to be in habitat where they can find prey and survive.

Meerkats usually eat insects in Africa or food they find underground. They also eat other animals like lizards, birds, and fruit. The insects they eat are scorpions, beetles and many more insects. 80% of a meerkat's diet is insects, which they eat everyday. When a meerkat finds something big that they could eat, it yells for help to kill it and eat it. As a mob, they work together to earn something big. When the mob gathers food they make sure they are safe and as well as their babies, so they wouldn't be harmed. When a meerkat finds food, they communicate and say they found food. They communicate by yelling at each other.

When meerkats communicate they use body language. They also use weird sounds to communicate. Meerkats work together in groups and they make noise if a predator is coming. When they find predators, they work in groups to protect themselves.

It would be hard to be a meerkat because predators are everywhere. That is why it is extra important for them to use each other and their features to survive. When a meerkat is in attack mode, their tails are up. So if you ever see a meerkat like that, stay away.

Elekat

Jonas Bagaoisan and Michael Tucceri

An elekat is a very interesting animal. Its appearance is unique as well as the way it searches for food and interacts with its own kind. There isn't any other animal in existence quite like the elekat.

The elekat has a striking appearance. With black stripes on its back and its tan fur, the elekat is a master of camoflague. It uses its coloring to blend into its surroundings. The elekat also has a long and furry tail that it uses to balance when it stands on its hind legs. An elekat normally stands on all fours, and is 3 feet tall from foot to shoulder. The trunk of the elekat can be as long as 12 inches, which it uses to slurp up water and food. The elekat has big elephant-like ears that it uses to hear predators approaching. Their appearance helps them survive in the habitat they live in.

The elekat's habitat is in the African plains. They make their home by using their trunk to dig a den in the ground. They live amongst the tall grasses on the plains. The grass makes it easy for them to camouflage. Their dens are also located near watering holes because they like to bathe in the water. Elekats live in large groups of 5 to 10 members, which is called a mob. The elekat has food surrounding their habitat, which means that they do not have to travel far to find food. Elekats choose locations that are heavily populated with fruit trees. They also pick the location of their den based on where the food is plentiful.

The elekat preys on and eats all types of food. For example, it eats tree leaves, grass, bugs, fruit and sap. 80% of the elekat's diet is made up of fruit, but if there isn't any fruit the elekat eats insects instead. The elekat uses its trunk to reach the fruit and leaves on the trees. They feed in their mob to help protect themselves from predators.

The elekat is vulnerable to a variety of predators. These might include rattlesnakes, red mongeese, alligators, lions, hyenas and even other elekats. Elekats stay in their mobs in order to protect themselves from these predators. If an elekat is caught alone by a predator it uses its trunk to make a trumpet-like sound to call its mob for help.

The elekat's intelligence level is high. The elekat is one of the only animals that can be trained to do human skills. For example, it can pick up a stick and scratch itself or use it as a fly swatter. They can also pick up small logs and dig with them. The elekat's intelligence, social behavior and appearance are extremely unique.

Corn Snakes
David Lopez

Corn snakes are a species of snakes that are mostly in captivity but a few do live in the wild. They are non-venomous and get their name by the way their stomach resembles a corn maize. They are a cold blooded reptiles which means their body temperature is the same as the environment's temperature around them. They grow to about four to six feet long. They are also the most non- aggressive and non-venomous snake out of all the snakes in the entire world.

Wild corn snakes live in overgrown fields, farms, trees, and abandoned buildings. They can be found in the states of Texas, New Jersey, Florida, and South California. In colder areas of the world, the corn snake hibernates during the winter. The corn snakes are less active during the winter, which means that they hunt less. When they are in the wild, they find a log or a small area underneath a rock to live in just in case of snow storms or predators trying to eat it while it sleeps. In captivity, the cornsnake usually lives in a terrarium with a heat pad or lamp due to the fact that the corn snake is cold blooded and that it needs a source of heat to survive.

The corn snake is a carnivorous animal. In the wild they prefer to eat mice and other small rodents, but sometimes they are found eating amphibians and other reptiles. In captivity, the owners of the snake feed them live or dead mice. Depending on their size, the corn snake gets different sized mice. In captivity, they are fed once a week but as they grow into an adult size in two to three years they are fed once every two weeks. The reason why they are not feed every week is because it eats the mouse whole.

The different colors of the corn snake vary in about 15 different colors such as orange, red, white, yellow, grey with black spots, orange with white spots, and a lot more different combinations of colors. When they first hatch from their egg, they are about six to twelve inches long and as they grow they can reach up to six feet in length and about one inch wide.

Corn snakes are like any other snake. They hunt their food by lunging at it and biting it. Some snakes inject venom from there teeth such as rattlesnakes and coral snakes. However, the corn snake does not have any venom so it bites its prey and then swallows it whole without chewing like any other snake. When you feed a corn snake in captivity and it is alive, it is important to watch it just in case the mouse tries to defend itself by scratching or biting the snake. In captivity, if you feed it a dead mouse it can eat its food without it fighting the snake. Even if it is dead, the cornsnake still lunges at the mouse and bites it. As the corn snake gets bigger you can start to feed it live mice without a problem because it can hunt easier.

In conclusion the corn snake is an interesting animal because of the way it lives without one of the most important tool of survival, limbs. Without limbs you could not run as fast away from predators or chase after your food pick up objects a lot easier, but snakes still find a way to adapt in the wild and captivity so that they can live without arms and legs they learned to slither then walk, sneak up on their prey and lunge rather than chase their prey, and they learned to use their mouths to grab things. These are just a few of the reasons a corn snake is such a fascinating creature.

Bobcats
Zoey Roman

There I was chasing a giant deer in the Rocky Mountains. My heart was racing and my face was rushing against the cold air. I was running 30 miles per hour. I was pretty fast but the deer was faster. It turned and jumped over a giant boulder that was 10 feet tall. I can only jump 4 to 8 feet, so I lost it.

Not catching the deer was a big deal because I only eat seven pounds of meat every month. So far, I only had three pounds. I was starting to starve and the winter was coming soon. I was very jealous of my cousins who live in the desert because they have a lot smaller and easier prey to catch. It was so hard to feed my kittens they might not survive this winter. We have a cave right where the deer herd is, but sometimes they don't come around.

Today was a long night because I am nocturnal. Sunrise was happening so we went in the dark cave and slept all day. It was sunset and I smelled food with my nose. My nose can smell prey and predators from several hundred feet away. Since we're not social and dont live with anyone we have to rely on ourselves. Luckily, we have an alarm system built in. The mates stay with us until the kittens are born and then they leave the den.

I left the den in search of the smell. As I was searching, I came across one of my predators the fox. I am very stealthy, so I snuck up behind a bush. My reddish brown fur and white underbelly helps me camouflage so its easier to hunt. I scared the fox away and I saved my kittens. The hardest part of being a mom of bobcats is they have multiple predators like the owls, coyotes, foxes, and eagles but we bobcats are born survivors and fighters and are very protective of our young.

I am one of the most fierce and aggressive cats in the animal kingdom. Us bobcats are very interesting and smart creatures.

BOBCORN
Zoey Roman and David Lopez

    The bobcorn is an interesting animal due to the fact that it has qualities of a snake and qualities of a bobcat. It is non venomous, which means that it is not poisonous. The bobcorn gets its name by the snake like head, tail, and body but cat like legs, ears, whiskers, and fur on the top of its body. The bobcorn's fur on top of its body helps it adapt in the snowy mountains.

    The bobcorn lives in the mountains of North America. It hides under rocks and logs in Alaska, California, New Jersey, Oregon, and Texas. The bobcorn is bulit for many different kinds of environments in the mountains. The kind of mountains it lives in are the snowy mountains in cold areas or grassy and rocky mountains in warmer areas. It has rough paws for the different kinds of terrain.

    The bobcorn has a big diet. It eats small animals and birds. Depending on the environment, they are in their eating habits vary. In the warm and colder mountains it eats birds, mice, lizards, and rabbits. In the colder mountains before it hibernates it eats a lot of food. The bobcorn hibernates for about two months, December and January. It awakens between February first and late February.

    The bobcorn is nicknamed the vampire because of how it hunts. It sneaks up on its prey then lunges at it jumping on top of it pinning it down with its claws then biting into its neck. The bobcorn then spits out the bones and puts them around their home to alert other animals that there is a bobcorn living there and that they should leave.

    The bobcorn has evolved to have fur on its body to protect itself from the cold weather in the mountains and to hide itself in the plants, snow, and rocks that are there. It has long and sharp claws to defend itself against predators and to help hunt its prey. The bobcorn is very long so it moves faster to catch its prey.

    The bobcorn's ears are pointed at the top like party hats that helps with their incredible hearing. It is a thin snake like body with ears that can hear over several hundred feet away. They have sharp retractable claws. There claws help them climb trees. Another reason they have these claws is for hunting so its easier for them to run and have a better grip on the ground. The bobcorn has a long tail it uses to help them balance. They also have whiskers that can alert them of their surroundings. They also use their whiskers to hunt at night to feel around and smell for different prey. They have sharp fangs that are non venomous that helps them bite into their prey, kill, and eat it.

    The bobcorn is the only cold-blooded mammal in the world. It is a land animal and is in the cat family but it is still a cousin of the snake family. This animal is interesting because it is related to both snakes and cats and that is fascinating and amazing.

Moon Jellyfish
Hali Saint-Lot

I was snorkeling in the clear waters in California by myself. There were so many fish of all colors to see! My other friends wanted to go look for tide pools because we were on vacation. I was okay with them being without me though because I am used to being by myself. I am an only child.

My parents were in our hotel room they said we were old enough to go places by ourselves. If I had any siblings I would want one with a similar name as mine, Sophie. I like my name, isn't it nice? Anyway, I am also fine with having no siblings and just being by myself.

While in the water, I saw clown fish, tuna fish, and fish I've never seen before! All of it was just amazing. I suddenly felt something tickling my foot, and I assumed it was a friendly fish. I turned around to look at it smiling when I saw that it didn't look like a fish it looked like a plastic bag! Then I thought to myself, I should take it out because it's polluting the water! So I did. The plastic bag squirmed in my hands. I was so surprised I dropped it back in the water. The plastic bag started drifting under the water almost as if it was swimming away. I thought to myself, this has to be an animal. I was curious, so I followed it softly kicking my flippers, so I wouldn't scare it. The animal went over and under small places, so I went around them. Finally, the creature stopped in front of a whole bunch of animals of the same species!
I was suddenly scared. What if the creature was dangerous? I inspected each animal carefully they all had stringy things under them. Hmm.. I thought to myself, interesting. I knew about most animals and where they lived and what they looked like, but why couldn't I figure out this one? I was in Monterey Bay and mom said we might see jellyfish, which is why I came out to snorkel… "Jellyfish!" I said while I was under the water. Oops! I thought to myself.

I came up to the surface and coughed out the water. The weird animals were jellyfish! I had found what I was looking for! " Hurray!" I said. Then I said out loud to myself, " What species are they though?" My dad also knew about animals that lived underwater, so I scooped some water into a jar then put the jellyfish in and went off to my hotel room to see my dad.

"Interesting jellyfish ya found, nice job Sophie!" said dad.

"Thanks dad! But actually, I came to ask what kind of jellyfish this was." I said.

"Oh okay, let me take a look at it," said dad.

I handed him the jar with the jellyfish in it. Dad inspected it then he rolled his swivel chair to a microscope he brought with him and put the jar under it. "It seems to look exactly like a moon jellyfish," he said.

"Great a moon jellyfish!" I said.

"And it seems to be 3 inches and it's a boy!" said dad while he adjusted his microscope. "It can be found in Monterey bay and the California Coast," said dad. "Their tentacles are small so they don't affect humans so that means it doesn't have nematocysts. Nematocysts is when a jellyfish has a lot of stinging cells in their tentacles, which is what makes them sting. Moon jellyfish are carnivorous and they eat things like small plankton and larva crabs. They can live up to 25 years. They also have for rings on top of their bell which is where they got their name from," said dad.

"Yay, I can keep it even when I'm an adult," I said. "And since it doesn't have nematocysts, I guess that's why it didn't sting me when I picked it up. I thought it was a plastic bag!" I said. Then my mind started wandering off. I thought about how cool it would be to keep the jellyfish as my pet and name it and put it in a tank and feed it and everything! "Man that would be cool!" I said.

"What would be cool?" asked dad.

"Well, I was thinking about keeping the jellyfish as my pet," I said.

"You would have to keep it in a tank with temperatures like -6 degrees C to 19 degrees C," dad said.

"Oh ,and your mom and I would have to have a conversation about that. Good idea though!" said dad.

"I'll ask her right now!" I said.

"Okay good luck!" said dad.

I ran to my mom and dad's room "Mom! Mom! Mom!" I yelled.

"Yes, sweetie," said my mom.

"I got a moon jellyfish and it's a boy and I wanna keep it as my pet, can I, can I, can I?" I yelled.

"Well, what did your father say? Let me come see it," said mom.

"Daddy said you and him have to talk about it, but can I please have it?" I yelled.

"Well, like your father said we have to talk," said mom.

"Aww..." I said glumly.

"That doesn't mean I said no."

"I know mommy it's just you know I hate waiting." Just then my friends Sarah, Teresa, and Tess burst into the door, like I said we were all on vacation together.

"Tess won a starfish as a prize for knocking down all the bottles in a carnival!" they screamed. Tess held a starfish in a tank grinning.

"Nice starfish Tess. My mom and dad are deciding whether this boy moon jellyfish can be my pet," I said. I picked up the jar with the jellyfish in it to show them.

"Oooooh it's so pretty!" they said all together, they did that a lot. Then they went to their hotel room.

The next morning, I woke up and creeped into the room with the jellyfish in the jar. I was interested in this moon jellyfish, it seemed curious when around me almost as if it wanted me, and me only, so did I which was funny. If I told my friends about this they'd laugh and tell me to snap out of it. Sometimes friends just don't understand you. You know what I mean? Well anyway the jellyfish seemed to be awake, I couldn't really tell it was just floating maybe it was asleep. I sat down in a swivel chair, crossed my legs and thought about a name for my jellyfish. So I started thinking. Maybe Bob? No, David? No, how about…. Charlie? That seemed like a good name to me. The moon jellyfish started to bob up and down. Charlie the boy moon jellyfish! I liked it.

I didn't want to wait all day for my parents and my friends to wake up, so I decided to go out, maybe for a swim, or a walk by the beach. I took a shower, brushed my teeth, and threw on some clothes, oh and I also left a note for my parents. I started to walk out to the beach. Maybe I'd find some more moon jellyfish if I went for a swim. I decided to go to the tidepools to see what my friends saw yesterday. I saw a lot of starfish and some other fish. All of it was boring to me. If it was a type of starfish I'd never seen before, it would be interesting. I like things that are unusual or new. Old things are just boring to me. I went back to the hotel room and my parents and friends were awake and having breakfast.

"Hi, Sophie!" said Teresa cheerfully.

"We saw your note," said mom.

"Come have breakfast with us," said dad.

"Okay," I said. I sat down and ate my breakfast. Today was a quiet morning. I wondered why. Maybe there was a surprise or something else. When everyone finished eating, they started to leave the table except my mom and dad. I was about to leave too when my parents told me to stay.

"We have some news for you Sophie," said dad.

"Good or bad?" I asked with a nervous face.

"Both," said dad.

"Good news first, you get to keep the jellyfish," said mom. "But you will be taking care of him."

"What about when I'm in school?" I asked worriedly.

"I'll take care of that," said mom.

"Now bad news," said dad. "We are going back to Florida."

"When?" I said.

"Tomorrow at six," mom said.

"P.m.?" I asked.

"A.m." said mom.

"Aww.. but that isn't fair. It seems like we barely got here," I whined sadly.

"I know time goes fast," said mom. Then my friends came in.

"Hey, Sophie wanna go out with us?" asked Sarah.

"Ya, come on we don't know where we're going, but want to come?" said Tess.

"Who would at this point?" I said annoyed and sadly at the same time. Then I rushed out of the room and went to the room I was staying in and slammed the door. Who did want to go? They'd find out soon we were leaving, and then they'd see, it's not fun anymore.

"What's her problem?" asked Sarah.

"Ya, we didn't do anything," said Teresa.

"Shes just a little upset right now," said Mom.

"Why?" asked Tess. She was sensitive when people were upset or sad or mad she felt like helping them.

"Well girls, I think it's time we told you too," said dad. "We're going back to Florida tomorrow at 6:00 a.m.

"What?" they said at the same time.

"Yep, sorry girls," said mom.

"It was so short!" said Sarah.

"Not really," said Teresa. Sarah looked at her once then ignored her.

"Well, I'm not going anywhere, anymore guys. I'm going to start packing up," said Tess.

"Me too," said Teresa.

"Guess, I should too," said Sarah.

The next morning, everyone was ready to go. We were all up early since our flight was at 6:00 a.m. We drove off to the airport.

"Well this is it…" said Tess.

"Guess it is," said Sarah. We had finished with our security checking and everything else and were about to get on the plane.

"Got your jellyfish?" asked dad.

"Yes," I answered. Well, I thought to myself this is sad, but at least I'm going back with Charlie the boy moon jellyfish!

Trumpeter Swans
Natalie Cote

I was gliding gracefully in the lake in my Canadian prairie. There are very few human disturbances and little pollution, that is why my parents like it here. I could tell my mother was watching me from the nest on the old beaver lodge my parents had found a long time ago. They worked together to build the comfy nest lined with feathers. I had five other siblings swimming out. Us trumpeter swans are really social. The rest of our flock was eating farther down in the lake.

"Don't go too far!" mom called. I was thinking about my family that moved to Alaska, some of them even live in the wooded areas and prairies of Northern United States. Us trumpeter swans can live in any of those places.

I looked out at the flock and I saw a small grayish whitish dot swimming towards me. I looked closer, of course it was Kleo. SPLASH!

"And she stops just in time!" Kleo screamed.

"Kleo!" I yelled. She just laughed.

"Hello! Whatcha want to do? I have lessons so I can't stay for long," she was speaking really fast.

"I have lessons too," I said.

"Oh, let's swim around," we swam around the pond for a couple minutes in silence. Neither of us could think of what to say.

"Kleo!" there was a call from behind us.

"Gotta go, bye!" she yelled and swam off. I had a while to myself, I remember that mom told my siblings and I that we would be learning about finding food!

A couple minutes later mom gathered us all around her for another motherly lesson, today, diving and finding food, I remembered! Sylvester, Kira, Rett, Mora, Calvin and me, Nela, gathered in a circle around her.

"Like so," she said as she dove into the water with her bottom sticking up. She was down there for a while but finally she came up. There were roots and other plants from the bottom of the lake in her mouth. "You want to dig in the dirt with your bill and find all the food you can!" She was squeezing the water out of the food like we all should, so her voice was muffled, but I could mostly tell what she was saying. "Really squeeze it!" she said again as she finished eating it up. "You want to be greedy, get as much food as you can! You want to get the roots and plants, but since you are all still cygnets or babies you may also eat worms and other aquatic invertebrates. As you get older, you will eat more plants."

"Do you ever eat fish or fish eggs?" Calvin asked.

"Very rarely," mom answered. "Do you want to try to find some food now?"

"Ya!" we all screamed. We went down the line, most had success, then it was my turn. I dove in and dug around, finally I got a grasp on some roots, I could feel the pressure of the water as I came back up. Everyone cheered me on. It was really fun learning how to dive. I wonder what daddy will talk about, possibly predators, if I remember correctly.

My dad made a loud call from the nest, it was a trumpet like call, that is where we get our name. We all swam toward the nest and got up onto the land. I felt almost awkward because our legs are set really far back on our body. We all gathered around my parents. My dad was as big as they get, he towered over us, 4 feet! He weighed 30 pounds and had a wingspan of almost 8 feet! My mom was about 25 pounds too, I know when I grow up I will be big because we are the largest waterfowl creatures in North America and the largest swans in the world! My parents can fly 40 to 80 miles per hour thanks to their hollow bones so they are not too heavy to fly. They are beautiful, they have the brightest white feathers and the darkest black bills. Right now, my feathers are only a light gray.

"Children, when you are 3 or 4 years old you will develop mating pairs that you may stay in for the rest of your life," he announced.

"How long is that?" Mora asked.

"You may all get up to 24 years old," he answered. "But our main lesson is how to keep safe!"

"Yay!" we all screamed. I knew it!

"When you were little eggs, we had to protect you from coyotes, river otters, raccoons, golden eagles, black bears, brown bears, gray wolves, ravens and wolverines!"

"Scary!" Sylvester said.

"Now you are threatened by all of them plus snapping turtles, minks, great horned owls and california gulls."

"Really scary!" I said.

"It's okay, that is why you have parents, to teach you how to survive, and to keep you safe while you are young."

"What are your threats daddy?" Kira asked.

"Well, your mother and I have to watch out for golden eagles, coyotes, red foxes and bobcats, but luckily we have wonderful adaptations to keep us safe. Our long necks help us find predators from far away so we can get away before they come."

"Our long necks are also helpful to trace food from far away as well," mom added.

"So we can do all of those things too?" Calvin asked in excitement.

"You sure can!" said a voice from behind us.

"Auntie Rosa!" we all screamed.

"Y'all want to learn about preening?" auntie Rosa asked. We all screamed in excitement! Everyone gathered around auntie Rosa.

"Preening is another way of saying grooming. It is a way to keep clean. Don't waste those quill worthy feathers, they make only the best of them!"

"How so?" Rett asked.

"Well back in the 1600's they would hunt us for our feathers because they make the finest quills, and they would hunt you little ones for meat. We were almost extinct!"

"Really?" Kira asked.

"Up until the 1800s, we are now illegal to hunt! But unfortunately there is sometimes illegal hunting, or we are mistaken for our relative the tundra swan, and they are not illegal to hunt."

"That reminds me, we are also threatened by man made structures, human disturbances, wire collisions and lead poisoning, so you have to be very careful," dad added.

"Oh yes, that is very true," auntie Rosa said. "But let's get back to our preening session. You use your bill and grab your feathers like so," she was almost combing them out with her bill. "Now you try!" we all started to preen, it was really easy and it felt good. "Good good!" she yelled.

"Now can we play?" I asked.

"Sure!" auntie Rosa said.

"Yes!" I swam out to my friend Kleo who also finished her lessons. "Kleo!" I yelled.

"Nela!" Kleo yelled back. We swam closer and closer to each other until we met right in the middle of the lake. I saw Rett and Calvin swimming around and laughing. "You wanna race?" she asked, Kleo loved competition.

"Sure!" I answered. We swam around until we found a straight clear path.

"Ready?"

"Set…"

"Go!" We were swimming as fast as we could, Kleo was in the lead, I was focusing on winning, when I suddenly saw a white something swim in front of the path, I tried to come to a stop but I couldn't. WHAM! I ran right into old papa Leonard.

"Whoa!" he yelled.

"I'm so sorry papa Leonard," I said nervously.

"It's okay but you know my eyesight is bad, it is getting dark, I didn't even see you," Kleo had just missed him, but she came swimming back.

"I won!" she yelled.

"That's not fair! I ran into papa Leonard!" I yelled back.

"Calm down it's just a race," papa Leonard said, he is very wise.

Suddenly dad made the loud trumpet like call and the whole flock came over including me and Kleo. "I still won," she said, I just laughed as we slowly swam towards my dad.

"It's getting late!" he yelled. "We should all go to sleep." We found a comfortable spot to sleep. I was next to my parents. I stretched my neck across my body, tucked my bill into my wing, closed my eyes and drifted off to sleep.

Moon Swan
Natalie Cote and Hali Saintlot

You may never have heard of it, but the moon swan is a beautiful animal that has moons on its back that glow underwater! That is where they get their name. Their appearance will surprise you in the next paragraph so have fun!

The moon swan is a very beautiful creature that has many different features. They have long thin tentacles as a tail that drags on the floor when they walk. There are four rings that look like moons on their back. On their head, there is a transparent veil or bell that covers and protects their sensitive eyes. The bell flaps and flows when they swim. Moon swans have a 4 inch long black bill and black eyes. These beautiful swans have long white feathery necks and feathers on their backs and stomachs. When they are babies, they are about 6 inches and as an adult they can grow up to be 2-4 feet tall. The male moon swans are slightly larger than the females. Their legs and webbed feet are transparent which is a way to hide and help trick their predators in the water where they live.

Moon swans live by water. They are scattered in the coastal regions all around the world. But to be more specific, they can be found in Japan, Europe and The Gulf of Mexico. They are mostly found in Monterey Bay and the California Coast. They live in large groups called smocks. They can swim in brackish waters, brackish water is a mixture of fresh and salt water. They make their nests in elevated places by the water, staying away from pollution and human disturbances is very important. They sleep clustered together in their nests with their smock. Most of their food is found at the bottom of the brackish waters.

Moon swans are omnivores. They eat, roots and plants from the bottom of the water. When they are swimming, they eat plankton, larva crabs, and fish eggs. They also eat mollusks, copepods, and crustaceans. When they eat roots and plants they squeeze the water out of their food before they eat it. They also eat snails, worms, and other aquatic invertebrates. They digest their food through eight canals before digestion is complete. Many of their abilities help them survive.

The moon swan may have many predators, but they have wonderful adaptations to help them survive. They're hunted for their luxurious feathers, and they are currently endangered. Their predators are sea turtles, large fish, shore birds, and larger land mammals. They are also threatened by man-made structures, lead poisoning, and wire collisions. They can fly to get away from predators but not very far or high. Another way to get away from land predators is to swim. Moon swans can stay underwater for a long time because they can breath underwater. The most surprising thing about that is that they don't have gills! Instead they use their specially designed nostrils on their beaks that filter the oxygen out of the water. Their tentacles don't harm humans, but they are used to sting and paralyze small marine life by simply touching them with their tails. Moon swans can withstand temperatures as low as -6 degrees C. Their long necks help them see predators from far away, so they can get away before they get to them. They also help them find food on land and underwater. They can live up to 20-25 years.

Overall, moon swans are very beautiful creatures that deserve to be recognized. The moon swan shouldn't be hunted, they should be protected by the law. Look, but do not harm these magnificent animals!

Mustangs
Roan Correa

Galloping across meadows, wild mustangs are a magnificent sight. Mustangs have been galloping across the meadows and fields of the U.S. for a very long time, but now their population is lessening due to domesticating. Another reason is cattle farmers, because they believe that mustangs take away grass that their cattle need. Soon wild mustangs may become extinct in the U.S., so if someone ever saw a herd of wild mustangs grazing in a meadow, they would be a very lucky person.

Seeing a lone mustang galloping across a meadow is a rare sight because mustangs always travel in large groups or herds. Some mustangs can form extremely strong bonds with other mustangs because of something like blindness. The horse that they become attached to becomes like a seeing eye horse. When young male horses become old enough, they break off from the herd they are in and either live alone or form their own herd.

Mustangs eat according to their surroundings. They eat mostly grass but will eat leaves, twigs, and even tree bark if they need to. They have also been known to eat bayberry twigs, rose hips, seaweed, and even poison ivy. Mustangs usually get their water from streams, rivers, ponds, and lakes. They also get their water from the grass they eat. Mustangs can go up to a whole week without water! Some of a mustangs' and horses' favorite treats are apples, carrots, and sugar cubes.

Mustangs come in a huge variety of colors. The colors include white, black, brown, appaloosa, which is white with small dark brown or black spots and a white mane, roan, which is a reddish brown, greyish brown, or dark brown with grey mixed in, buckskin, which is a light yellowish or reddish tan with a black mane, and chestnut, which is a reddish brown with a reddish mane. They are shorter and stockier than regular horses and in the winter they grow a long shaggy coat to keep them warm. In the spring, they lose their winter coat and become sleek and shiny. Adult stallions can weigh up to 450 pounds. They are an average of 16 hands (a hand is about 4 inches), they can be as long as 6 feet, and their average lifespan is 20 years.

Mustangs have been roaming the plains of the U.S. for many years. The first mustangs came from Iberian horses. The Iberian horses were brought to Florida and Mexico by European settlers. Some of these horses escaped or were captured by native americans and spread rapidly throughout the U.S. Soon, horses replaced dogs as carriage pullers and greatly increased success in hunting, especially bison hunting. Cowboys captured mustangs that roamed the great plains from the 18th century to the 20th century, to use for cutting cattle and trading.

Due to domestication, there are not many mustangs left in the wild. There are many organizations trying to restore this huge part of american history. Maybe soon mustangs will be running free through open plains again.

All About Porcupines
Pierce Maginnis

Its very hard to believe that a porcupine has about 30,000 quills. Thats just one of a porcupine's cool facts. There are about 24 species of porcupines in the world. A porcupine's Latin name means "quill pig". Porcupines are also the third largest rodents in the world following the capybara and the beaver. They are about 3ft 1in long.

A porcupine may look nice and cute at first look, but don't go petting them! The fur of a porcupine is usually black, yellowish-brown, gray, and very rarely white. When they grow up, they weigh 5-18 pounds while others can weigh up to 25 pounds. They are mammals/endothermic so they can regulate their own body heat. They have claws that are 1 and ¼ inches long and a set of five toes on each foot. On a porcupine's palm, they have a pebbly surface without fur.

Porcupines have a lot of life-saving adaptations, like their acrobatic tails. Some porcupines have tails that greatly help them maneuver through trees, while others can barely use their tails. Porcupines are nocturnal which means they hunt at night. If you have ever heard rumors of a porcupine shooting their quills at someone, it was not true. Porcupines cannot physically shoot their quills at anything. Now don't go thinking these quills are an inch long, the quills on Africa's crested porcupine are nearly a foot long, but they can also be very soft. Baby porcupines have soft quills at birth, which harden within a few days of birth. They are a good swimmers with their hollow quills keeping them afloat, and they don't hibernate. They have a lot of ways to communicate, including moans, grunts, coughs, wails, whines, shrieks and tooth clicking.

A lot of people don't know that porcupines don't eat meat. Porcupines are herbivores. They eat about a pound of food a day. They can eat leaves, twigs, plants, fruit, springtime buds, and the layer of tissue beneath the bark of trees when any of their other food sources aren't available. Their food sources usually deplete around winter, so that is when they resort to the bark on trees.

Porcupines spend most of their time alone, but in the winter, they are a little more social. In the winter, porcupines will sometimes share dens in caves, decaying logs, and hollow trees, and find food together. They have great defenses, so they aren't afraid of animals hunting them down when they are alone. Males and female porcupines are territorial, although men defend their territory more than females do. They won't go far from their dens unless they need to get food.

Porcupines can live in so many different places! North America, South America, burrows in southern Europe, Africa, and southeastern Asia. They can also live in desert shrublands. Some live in the chaparral, while others in mountains, grasslands, woodlands, sagebrush, swamps, tundras, and thick forests.

Porcupines can do so many cool things from tree-climbing tails, to having 1 foot long quills that they can use to jump from tree to tree and fend off predators. Next time you see a porcupine, remember all of their cool adaptations and tricks that they can use.

The Porctang
Pierce Maginnis and Roan Correa

The Porctang, which is a rare and magnificent creature, is a wonderful and sometimes frightening sight. With their sharp spines and mustang like features, these creatures are only dangerous when threatened. To see one of these rare animals is an honor.

Porctangs live in so many different places! They live in North America and some parts of Central America. They are mainly located in Canada and remote- uninhabited parts of the U.S. Their home consists of a large hole filled with soft grass and leaves. The herd will stay in a group of nests for up to one year, then they will move on to a new group of nests. They build these large nests in thick forests for protection against predators. Their nests are always at least 20 feet away from each other. In the winter when there's only snow and no grass, they eat the dry grass from their nests to keep them alive until the spring. In the spring, the porctangs' nests are all gone so there is nothing to clean up. Also, in the spring, the porctangs' predators are well and active so they must be extra careful.

Porctangs are only dangerous when threatened. At birth, a porctang has soft quills on its neck and tail, which harden within a few days. If needed, they can shoot their quills on their neck and tail up to four feet away to protect themselves. The spines have barbs which make it very difficult to remove from an animal's skin once they are stuck. A porctang's spines lay flat on their mane and tail. When they get frightened or scared, their spines stick up to protect themselves. The porctangs main predator is a mountain lion.

The porctang has a lot of life saving features. They have hooves which they will use to kick out with and fend off predators. Porctangs are mammals, so they can regulate their own body temperature. The remarkable part is their tongue, which can easily be projected up to 50% of their body length, and sometimes up to 75%. Their tongue has a lot of muscles, which when suddenly contracted combine to instantly shoot out the tongue. It's very useful when they have pesky horse flies flying around and bugging them. The tip of the tongue, which is relatively large and club shaped, is covered in a sticky substance that can stick to almost anything. Thanks to the tongue's elastic qualities, it can be repeated immediately if needed. Their tongue recoils into their throat when they are eating, so it's not a hazard.

Porctangs just love eating. They are omnivores, so they eat mostly grass but will eat leaves, twigs, tree bark and even insects when grass becomes scarce. When they are near an ocean, they like to snack on seaweed found near the shores of oceans. They usually get their water from streams, rivers, ponds, or lakes. When they can't find a stream, river, pond or a lake they can live up to a week without water. This is because they get a lot of water from the grass they eat. When they can find a ranch or farm, their favorite treats are apples that they take from the orchards.

With all of these amazing, live saving features, Porctangs are one of the most strange and interesting creatures on earth. If you ever see one, know that you are a very lucky person.

Peregrine Falcon
Justin DiProfio

In the sky there is a cluster of pigeons, however in the middle of the cluster sits the incredible peregrine falcon. Peregrine falcons are not that hard to spot. They can be found worldwide except for in rainforests and cold, dry arctic regions. They prefer open habitats such as grasslands, tundras, and meadows. Sometimes a peregrine falcon will live in an urban city with a lot of tall buildings.

Peregrine falcons are easy to spot because of their striking features. Adults have blue and gray wings, dark brown backs, buff colored undersides with brown spots, and white faces with black stripes on their cheeks. They have hooked beaks and strong talons. Peregrine falcons have pointed wings with a wingspan ranging between 74 to 120 cm. Their tail length is 13 to 19 cm and their body length is 34 to 58 cm.

Peregrine falcons have an incredible way of hunting. Peregrines usually hunt with either a swift chase or a fast dive. Peregrine falcons sneak up on their prey and then dive at them at over 200 miles per hour. The peregrine hunts by flying high up in the sky. It then dives down super fast making its talons into a fist. The peregrine falcon then hits its prey in the back of the neck paralyzing it and then eats the paralyzed animal.

Peregrine falcons eat primarily birds. These birds include songbirds, ducks, and even bats. Starlings, pigeons, and doves are their favorite meals. Aside from birds, the peregrine falcon also eats small mammals, large insects, reptiles, and fish. What they eat depends on where they are. If they are in a grassland, they will probably eat rodents where as if they live in an urban city it's likely that they will eat pigeons.

The peregrine falcon is an amazing bird with some very cool features. Living in many habitats, eating many things, hunting in a very impressive style, and having many striking features makes the peregrine falcon an outstanding bird.

Gouldian Finch
Elijah Devillanueva

The gouldian finch is a very colorful and unique animal. It has a lot of different colors on its body. The most noticeable, is the red and yellow. They have white beaks and a yellowish belly. The colors vary depending on the bird and if it is male or female.

Wild gouldian finches eat mostly seeds. They can also be found looking around for bugs and berries. When they are lacking something in their body, they crave the food that will help. Gouldian finches that have been kept as pets can sometimes eat lettuce and sliced grapes as well.

Gouldian finches live wherever they can find the most food. They can be found in places with a lot of grass and trees. The most common areas they live in are New Guinea, New Australia and their surrounding islands.

When these animals are babies, they are not the same as when they are adults. Not only are the babies much smaller, they are also born without feathers. When the feathers do grow in, they are not as colorful as the adult feathers. The chicks also have light reflecting bumps at the end of their beaks to help their mother find them in the dark.

Gouldian finches are not that common, but they are still out there somewhere. There might not be many because they have a lot of predators who eat them and their eggs. These include small hawks, snakes and some lizards. In some places there are some animals that harm them because of where the gouldian finches happen to live. One of these predators is the house cat.

Gouldian Falcon
Justin Diprofio and Elijah Dellanueva

In New Australia there are many marvelous birds, however, deep in the forest habitat there nests the most marvelous bird of all. It can be mistaken for the gouldian finch, a bird that lives in the same area as the gouldian falcon and has the same colors as it. If you look close, however, you will see that the gouldian falcon looks very distinct.

The gouldian falcon is easy to spot because of its elaborate colors. Gouldian falcons usually have a bright yellow body, a whitish curved beak, a red or black face and a purple breast. Their wingspan is 70 to 120 cm, their tail length is 13 to 19 cm, and their body length is 34 to 58 cm. Adults have blue and gray wings. There are 19 types of gouldian falcons worldwide each type eating a wide variety of foods.

The gouldian falcon eats many things. They primarily eat birds such as songbirds, ducks, and bats. Starlings, pigeons, and doves are their favorite meals. Aside from birds, the gouldian falcon also eats small mammals, large insects, reptiles, and fish. They get most of their water out of the animals they eat but, they also can be found drinking water out of a nearby stream. The gouldian falcon does not only hunt other animals for food. It also eats grass seeds, fruit and oyster shells. These foods are available in the gouldian falcon's habitat.

The gouldian falcon is not very common, because it only lives in one place. This place is New Australia. It prefers open habitats like plains, grasslands and meadows. It likes these habitats because it makes it easier to hunt. They can also be found in forests, tundras, or big cities hunting pigeons, which is fairly easy due to its amazing hunting skills.

The gouldian falcon is famous for its incredible hunting skills. It can hover over its prey stalking it. Then, when its prey begins to suspect something, the gouldian falcon dives down on it at 320km making it the fastest animal alive. Before its prey turns around, the gouldian falcon hits it in the back of the neck paralyzing it. It can attract prey with its colors. Its colors also help it blend in with its surroundings.

The gouldian falcon is an incredible bird. Being very rare, living in one place eating many things, and breaking the record for the fastest animal the gouldian falcon is an outstanding bird.

Polar Bears
Hailey Pryor

Polar bears are big, soft, and incredible mammals. They are only considered marine mammals because it depends on its marine environment for survival. There are such neat animals and really aren't as vicious as they seem.

Polar bears are like big, fuzzy stuffed animals. Polar bears have thick white fur and thick blubber underneath to protect them from the cold. They also have longs necks and narrow skulls. Polar bears have four large paws. The front paws have toes on them that are partially webbed that help them swim. They also have sharp claws that grip the ice so they don't slip and fall. Polar bears have a large nose that has an amazing sense of smell that it uses to track down dinner. They also have nice sharp teeth that they use to tear into prey and kill it and eat it. Polar bears can weigh for a male between 550-1,700 pounds and for a female between 200-700 pounds. They are also be up to 8-10ft in height , when they're standing up. Polar bears also have great camouflage because their snow white fur blends in with their environment around them. Polar bears spend all winter hunting on ice and looking for food to stalk up for denning and breeding.

Polar bears live in the frigid, harsh, and icy areas of the world. They live in the Arctic areas that are surrounded by freezing cold, icy open water. They use the icebergs as platforms for hunting and getting around. Polar bears can also be found in parts of Canada and Alaska in the icy, colder areas. Polar bears can't travel far because their habitats aren't very big. Canada is one of the 5 polar bear nations along with the United States, (Alaska), Russia, Denmark, (Greenland), and Norway.

Polar bears are the most carnivorous bear in the bear family. They mainly eat seals, birds, kelp, walrus, narwhals, beluga whales, and bearded seals. They can go up to 10 days without finding any food because they can conserve the energy and store it to use later. That's a long time to go without food. Thier diet is full of other animals, too.

Polar bears ability to camouflage to make hunting a lot easier so they can easily say hello dinner. Polar bears have an amazing sense of smell and can smell prey from up to 20 miles away! Polar bears have fur that blends in with the ice which gives them great camouflage which is very helpful if you're a big animal that needs something to eat. They are big carnivores and eat large animals such as seals, narwhals, and beluga whales. One of their very top notch hunting strategies is called still-hunt. They will literally sit on some ice for hours at a time and wait for the prey to come up for air. They will spring into action for attack. What's funny is even though polar bears live in a freezing cold climate they can actually believe it or not get too hot if they are moving quickly. Which is why they move kind of slow because they don't want to overheat in the snow.

Polar bear cubs are small, curios, white bundles of joy that are ready to explore the outside world. When the polar bear cubs are born they are only about one pound and can't see or hear anything because their eyes and ears are sealed shut. After about two months their ears and eyes open and they are ready to explore the outside world. When they leave their den the cubs stay close to their mother. While they are taught how to survive in the arctic. After a few years the cubs learn things like hunting and swimming. Cubs born from the same litter are great playmates. They restel. fight, play, and chance each other, Once they are about two and a half they are sent out into the open world to live but, they sometimes will stay for two more years if they decide to do so.

Even though polar bears aren't endangered they still face many threats like global warming, melting ice, poisonous pollution, and oil collecting. To most animals humans are a big threat but, to polar bears they are not. Scientist say this because they live so far away from us in such a cold climate that we just don't bother them. They say there are about 25,00 polar bears living today. Global warming is a big problem because it melt the ice and the land they have for living and life. So when the ice melts it causes big pieces of ice to break away making their home smaller little by little. The oil collection is a problem because the natural oil that companies sell is found underground in the arctic. The companies set up large mining rings that collect the oil making really loud noises and causing natural disasters and oil spills. The loud noise will also frighten a polar bears with the awful sounds. The poisonous pollution is a problem because it releases toxic fumes causing animals to get sick so when the polar bears eat the animals the get sick too. Spreading the toxic fumes to the other animals is really affecting them but, they aren't endangered, yet. And honestly they need to eat too! It would really be a shame to loose them since they are such incredible creatures and their are so many things we can learn from them if we just pay close attention.

Clouded Leopard
Sophia Cheney

The clouded leopard is one of the smallest cats in the animal kingdom. They are only slightly bigger than a house cat. They mostly live high up in the trees for protection and to sleep. Their scientific name is Neofelis nebulosa and they're classified as mammals.

The clouded leopard can live in many areas. The areas that they live in are all on the other side of the world. They can be found in areas such as Nepal, Borneo, Myanmar, Thailand, Southern China, Vietnam, and Indonesia. Even though the leopard can be found in these areas, they are hard to find because they live deep in the jungle and spend more time in the trees than on land. Trees are one part of their habitat.

Their habitat is where they live and survive. They can blend into their habitat. They live in the trees, dry woodlands, tropicals, and in the rainforests. The clouded leopard lives in these types of conditions. They learn how to survive in this habitat. They can protect themselves by staying up in the trees to keep away from predators. They also can spot their prey for hunting.

Hunting is what helps the animal survive and not starve to death. The clouded leopard eats a lot of food which gives them energy. They are very stealth hunters. They hunt monkeys, wild pigs, squirrels, and birds. They hunt alone instead of hunting with a pack. They eat meat and that means they are carnivores. They eat squirrels, birds, wild pigs, and monkeys to get protein for their hunting energy. They have an average amount of energy to hunt for food. They hunt in the trees and the ground. They can get into small places because of their size.

The clouded leopard is a small animal. They are the smallest type of cats with dark brown spots that have black outlines. Their physical appearance helps them camouflage. The size of a clouded leopard is three feet long. The tail is also three feet long. From their shoulders down is ten-sixteen inches long. The males are larger than female leopards. They live up to seventeen years. During their life, they reproduce and have kittens.

A clouded leopard can have two or more kittens at a time. The litter number can be high or low. The kittens open their eyes at twelve days old. They weigh three to six ounces. They are nursed until they are five months old. They get their parents coloration at six months old.

Their name came from their coloration and markings . They got their name by living high up in the trees and the clouds. They have large spots on their body and good camouflage when they don't want to be found. They are the best climbers in the jungle and very interesting animals.

Clouded Bear
Hailey Pryor and Sophia Cheney

Clouded bears are fuzzy, spotted, amazing mammals. They are gentle and are not a threat to humans. They are truly incredible creatures and can teach us things we might never know with out them.

Clouded bears have really soft fur that has have big black and brown spots. They also have sharp claws and a large nose that has an amazing sense of smell that it uses to track down prey. Their nose can smell something up to 20 miles away! They have sharp teeth that they use for hunting and eating. They also have long spotted tails and have four large paws. The front ones have toes on them that are partially webbed so that helps them swim. Clouded bears are 4-5 feet long and 7 feet when standing up right. They also have big eyes that give them great eye sight to see the world around them.

They can only be found on the continents of North America and Antarctica. They live in the woodland areas and forests in Alaska and the icy terrain of Antarctica. The trees make a good place to run. The areas are surrounded by freezing cold, icy open water. There is a lot of snow where they live but in the spring and summer it gets warmer and sometimes can rain. When they hunt, they use the icebergs as a platforms or stepping stones for getting around on the water. They will often hang out in the forest or woodland areas of Alaska to keep away from humans and they also love the trees for sleeping or hanging out in them. In Antarctica, they stay on the ice unless they're hunting and looking for food then they will swim in the water.

They eat meat to survive in the wild. They happen to be on of the most carnivorous bears in the bear family. They eat birds, seals, squirrels, salmon, and walrus. They can go up to 10 days without food because they can conserve the energy and store for times of need. They hunt alone in the wild and when they do have young, they will occasionally hunt with them.

They spend all winter hunting on ice and looking for food to stock up for denning and breeding. One of their top notch hunting strategies is called still-hunt. They are still the entire time until prey comes. They will sit on the ice for hours at a time and wait for the prey to come up for air when they will spring into action to attack. The clouded bear has fur that blends in with the ice which gives them great camouflage which is very helpful if you're a big animal that needs something to eat. What's funny is even though clouded bears live in a freezing cold climate they can actually, believe it or not they get too hot if they are moving quickly.

Mothers can have two or more cubs in one litter. They weigh three -six ounces to one pound when they are born. After about two months, their ears and eyes open and they are ready to explore the outside world. They are nursed until they are five months old and at six months old they start having their adult coloration. After a few years, the cubs learn things like hunting and swimming. They stay close to their mother while they are taught how to survive in the Alaska. When the cubs turn two they are sent to survive in the wild one their own, however, they may stay up to two more years with their mother until four if they please.

Even though human aren't a big predator they still face many threats. Global warming is a big problem because it melts the ice and the land they have for living and life. So when the ice melts it causes big pieces of ice to break away making their home smaller little by little. The oil companies collect natural oil that is found underground in Antarctica and they sell it for profit. The companies set up large mining rings that collect the oil making really loud noises and causing natural disasters and oil spills. The loud noise from the machines will also frighten the clouded bear. The poisonous pollution is a problem because it releases toxic fumes causing animals to get sick, so when the clouded bears eat the animals the get sick too.

It may seem weird that humans that aren't a big threat to the clouded bear, but scientists say this because they live in colder parts of both places and we just don't bother them. Clouded bears are incredible and rare animals and would be such a shame to lose.

Red-tailed Hawk
River Stiles

    The red-tailed hawks are fast and skilled flying birds. They can be spotted all over the continent of North America. Their red and rusty brown color makes them easy to identify. They can soar high up in the sky or dive down to land when they need to. They are fascinating birds.

    The red-tailed hawk has many features, like their sharp-hooked beaks and talons. For flying, they have long broad wings. Also, their bones are hollow, so they're lighter. In their bodies they have nine air sacs, that take up the room in between their organs which makes them even lighter! Red-tailed hawks can see you before you see them, their eyesight can see over a mile away! When you do see them they appear rusty red.

    Red-tailed hawks eat small mammals and birds. They can hunt and catch mice, lizards, birds, rats, insects, and even roadkill, if they're hungry enough. They then take it back to the nest, or they eat it right away. They eat by holding down the prey and tear at the flesh.

    Red-tailed hawks live in various habitats. Such as grasslands, woodlands, prairies, deserts, plains, or even the city. You can see them all over North America. Wherever there's prey and high places for nests, there's going to be red-tailed hawks.

    Red-tailed hawks soar over large fields to look for prey. When it finds one it goes into a dive over 200 mph. It then sticks its talons out and then grabs its prey. They also perch on high branches to find prey. Red-tailed hawks can also hunt in packs, one will chase it and another will dive for it.

    Red-tailed hawks live in pairs. Only adults males have red tails to attract the females. Before mating, they do a dance in the air that ends in a dive. A chick has a egg tooth that helps it break the egg, then the tooth snaps off. Chicks get a red tail when they become an adult. They live for about 17 years in the wild.

Red-tailed hawks are amazing creatures and should always be admired. So the next time you look up and see one, think about what they do and how they do it.

The Cheetah
Alex Alvarez-Milch

There I was in the tall grass, in the Savanna, waiting for the right time to pounce on my fresh prey, the leader impala, and the chase was on. I was going as fast as I could, which is more than 90 mph, but then the impala's speed increased more every second of the chase. I jumped, so I could capture the impala, but it made a sharp turn right at that same exact moment. I turned right in midair and when I landed, I had my paws in a tar lake. At that exact same moment, the impala got a chance to run away.

When the chase was over, I ran back to the waterhole disappointed of my epic failure. I traveled from the tall grass to the waterhole. When I got to the waterhole I saw some elephants and giraffes passing by. I also saw some zebras and gazelles grazing in the grass. On the other side of the waterhole, my coalition was waiting for me to serve them lunch. When I got to the other side, the leader, my father, thought that I had triumphed over the impala. When he found out that I failed, he went back to the tall grass and waited. I thought, I did well since I am only 6 years old but my father did not.

My father was an old man, at 12 years old that is a cheetah's lifespan. Since he was the leader of the coalition, he had to watch over us. I always looked up to my father like he was an idol to me. When I was little, he always had bright black spots and golden fur. My mother, on the other hand, was black on the bottom and white on the top of her body.

When I got home in the tall grass, and ate rabbit for dinner, I realized that my father was not going to be leader anymore because he was too old. I was going to have to take his place. So I told my cub sisters and brothers and my mom that he was not leader anymore. After realizing it, I went to sleep knowing that I was the new leader of the coalition and I would have big responsibilities to fulfill.

The Chawkee
River Stiles & Alex Alvarez-Milch

By The Pond...
   There I was, perched on a tree branch, watching a nearby chawkee chase a rabbit, triumph, then eat it. He then flew onto the branch that I was perched on and he said to me, " How about we go hunting, I bet you are hungry."
   "No thanks, I'm good," I replied, ignoring the hunger that was inside of me.
   "Are you O.K? he asked listening to the growling in my stomach.
   " Well, if you need me I'll be with my mother eating dinner," i said.
   " i thought you weren't hungry." he replied.
   " what? i can't hear you!" i said gliding off into the sunset.
   I'll' never understand that kid, my father thought to himself following me back to our tree.

In The Air...
   Once in the air, I soared all across the african savanna, the grasslands and over herds of deer, zebra, and wild buffalo. once i turn 16 years old i get to have my own tree to myself. on my way home, i also saw some really cool trees that in a couple years that could be mine.
   Chawkees are a mixture of a red-tailed hawk and a cheetah. my bones are hollow so it is easier to fly and glide. i also have some very sharp talons for good landing gear and it is easier to catch prey. I have a long beak for more room so i can have a lot of prey in my beak at once. i have long wings so i can glide longer and faster. chawkees can fly at speeds from 200-280 MPH.

Back At The Tree...
   I sat next to my mother eating the antelope she caught. My dad flew on the branch and started talking.
   "Ralax, you know that one day you'll take care of the pack." he said with a long sigh.
   "I know that dad, you told me 10 seconds ago during our flight home." I replied with annoyance from my father who kept repeating and repeating it over and over again.
   "You will have to prepare for it, with great jobs comes…" My father started to say.
   "Comes great responsibility, I know that dad you told me just yesterday!" i replied with anger. the next day ilwoke up knowing today was the day I was the new pack leader and we started off the day by hunting.

Over The Fields…
   I was flying high in the sky when i saw a roamer just 1 mile away from the pack. The roamer was a big creature with hard skin, what came out of the it was walking right towards the pack.
"run!" I yelled at the pack then they, Including me, ran to a near by tree hole and we found out we were running away from a walker with a boomstick. We knew we could not fly, because they would shoot us. We thought it was the end of us but a herd of buffalo came in and they took the buffalo instead, because everybody knows that buffalo are bigger than chawkees, so we were saved!

Back In The Grasslands...
   We were talking about my dad and how great he was and when he swooped in with his big wings. he wanted to talk to me about my first day of leading the pack. I explained about the buffalo herd and the boomsticks and he was amazed and proud of my quick-thinking and bravery. then after the amazing story, he asked me one final question.
   "Wanna go hunting?"
   "Yes" I replied with joy.

THE END
or is it?

Giant Panda
Maddie Kreis

The giant panda presents itself as a cute and innocent little bear, but giant pandas can be as dangerous as any other bear. One interesting fact about them are that they used to be meat-eaters but not anymore. They used to be like any other bear. They prefer to eat bamboo now. Let me tell you more.

The giant panda is covered with black and white fur all over his body. Some parts where its body is has black spots are around his eyes, ears, legs, and its big, bulky muzzle. Then the rest of the giant panda's body is white. Giant pandas have soft, thick fur to keep them warm in the freezing forests. Giant pandas can be as long as two to three feet when standing on all four legs. The pandas in captivity live up to be twenty to thirty years old. The pandas that live in the wild live between fifteen to twenty years old.

Most people think that pandas only eat bamboo, but they eat a lot of other things too. To start off, pandas eat bamboo, flowers, vines, honey, rodents, tufted grasses, and green corn. Bamboo is 99% of its diet, it's pretty much what it eats. In one day, a giant panda has to eat twenty three to thirty six pounds! That's a lot. Giant pandas have a strong jaws, large molars. and use their thumbs to help them eat tough bamboo. They use those three things for when they are eating bamboo because bamboo is pretty hard to tear through.

Giant pandas don't really live close to where we are.. Giant pandas live in a few mountain ranges in central China. In the mountain, they live where there is a big forest filled with lots of things they like to eat. They prefer to live in forest where it is misty, damp, and where it has high altitudes. They like to live where there is very thin air. When pandas sleep, they especially like to sleep in dens. Once every three years the panda finds a new den.

When a giant panda is born, it is the size of a rat. When a giant panda is six to eight weeks old, it first gains sight. Then at four months old a panda first learns how to walk. Soon after that, a panda goes on it's own at eighteen months, sometimes baby pandas stay with it's parents for three years. By three years a panda is fully grown. At that age, the panda will soon leave and find it's own habitat to live in.

The giant panda is a very exciting and an interesting animal. They have different ways to eat and have special places to live. The giant panda is very unique because its kind of not like any other bear. Most bears catch fish, but they eat bamboo. The giant pandas are very different in many ways.

Wombat
Eden Pardo

    In the Australian outback, the wombat burrows underground to escape the hot sun. This creature spends its time taking care of its young, searching for food, and protecting itself from predators. From far away it may appear social and gentle, but it is actually very powerful. It uses its strong features to survive in the animal kingdom.

    The wombat has a distinctive appearance. They have rough and soft fur on their bodies. A wombat has long rodent like teeth. They also have long sharp claws on their paws and can weigh between 44-77 lbs and can live up to around 30 years old. They have a strong sense of hearing and smell. They are usually 1-2 meters in length.

    Even though wombats look cute and cuddly they can also be very protective and aggressive. Wombats never let an intruder come into their home. Their predators are the tasmanian devil and the dingo. They have a pouch like a kangaroo, but theirs is backwards. The reason it's backwards is so that when the wombat digs into the dirt it doesn't get inside it.

    Sometimes it can take a wombat 14 days to fully digest their food. The reason why it takes them so long to digest is because they need to be able to extract the maximum amount of nutrients from the little amount of food they eat. They are herbivores and only eat roots, grasses, bark, and herbs. They wander outside at sunset to start feeding because they are half nocturnal.

    Wombats live in Australia in forests. They live in burrows that they dug themselves with their sharp claws. A wombat's burrow will stay comfortable at 78 degrees fahrenheit when it is 105 degrees fahrenheit outside. Their burrows that are dug a few feet below the ground helps to keep the wombat nice and cool. They cover the floors of their burrow with soft and dry grass. Some wombats are social and some keep to themselves.

    Wombats can have babies at the age of two. They only have one or two babies at a time. The baby stays in its mother's pouch for about 7-10 months before being born. They use their backwards pouch to hold their babies.

    Wombats are very common in Australia. They are amazing creatures because they have adaptations that help them survive in extreme climate and situations. Don't be fooled by their appearance they actually have a strong will to survive.

The Giant Wompabat
Eden & Maddie

    The giant wompabat is an amazing creature. Giant wompabats are creative, interesting, and one of a kind. They have some techniques that are different from other animals. Some their features are for helping them eat. Sometimes the giant wompabat is known for looking like a koala.

    The giant wompabat has gray fur around its eyes, ears, legs, shoulders, and muzzle. The rest of their body is white. They have a pouch like a kangaroo to hold their young. They have long sharp claws and their hearing and smell are very strong. They can weigh 44-77 Lbs and are 1-2 meters in length. Most giant wompabats can live up to 27 years old; the reason they live that long is because they have a healthy diet.

    Giant wompabats are carnivores and herbivores which means they are omnivores. They eat roots, bamboo, flowers, vines, tufted grasses and small rodents, but they prefer to eat more of bamboo and roots. Sometimes, giant wompabats take up to 14 days to fully digest all of their food. The reason for this is because they need to be able to extract the maximum amount of nutrients from the plants. They eat 10 pounds of food a day. When they eat their food, they have some special parts to help them. They have a strong jaw and large molars to help them chew. They also have a strong thumb to push down the tough bamboo and rip it in half. Their food can be found everywhere in their home.

    The giant wompabat lives in a few mountain ranges in Malaysia. They also live in the Malaysian rainforests. They live in dens and sometimes sleep in trees. They cover the floors of their den with soft and dry grass. Some of them can be social animals, but they can be very protective of their home. Once they are three years old, they leave home in search of their own. They like being in the misty and rainy mountains, where their one predator lives.

    The giant wompabat presents itself as a cute and cuddly mammal but they can be very protective and aggressive. Their predator is the ledingard, a species that related to the leopard and the dingo. They use their long sharp claws to defend themselves when needed. To save energy, the giant wompabat can slow down its heart rate while napping. For the cold, misty forests, they have big, heavy coat to keep them warm. Their large muscular figure helps them when telling others to stay away. They spend most of their day trying to find food.

    The giant wompabat is a very rare species. There are not many giant wompabats left. In the big forest, giant wompabats are hard to find. So, anyone who sees a giant wompabat they are really lucky. Everyone should take the time to learn more about them and discover their interesting features.

Hammerhead Sharks
Miz Valverde

Hammerheads are one of the most unique looking species of shark. They have a flat-wide head shaped like a hammer. Their eyes are positioned on the sides of their head, giving them 360-degree vision. Hammerhead sharks have extremely sharp teeth that are serrated, like the blade of a knife. They are usually gray on top and white on the bottom. The largest of the hammerhead sharks species can weigh 500 pounds and measure up to 13 feet.

Hammerhead sharks can be found all over the world in both shallow and deep tropical waters. They can mostly be found right off the coast, but some have even been spotted far offshore. Schools of hammerheads have been seen near Malpelo Island near Colombia, Cocos Island off of Costa Rica, and near Molokai in Hawaii. Large schools have also been spotted off the coast of south and east Africa. In summer, the hammerhead sharks find cooler water.

Hammerhead sharks eat a large variety of sea creatures. They eat bony fishes, crabs, squid, lobsters, stingrays, and other sea creatures. They use their large heads to trap stingrays against the ocean floor. They can go without eating for a few weeks because they can store extra energy from fat. Hammerhead sharks swim in a school during the day but prefer to hunt alone at night. They have eyes that let them scan more quickly than any other shark which lets them find more fish. Hammerhead sharks have special sensors on their heads that help them sense the electrical signals that ocean creatures give off. Some Hammerhead sharks have been known to eat their own kind.

Hammerhead sharks mate only once a year. The male shark bites the female violently until the female agrees to mate with him. Hammerhead sharks do not lay eggs like many other fish, they lay live babies. A litter of Hammerhead sharks can range between 6 and 50 pups. Baby Hammerheads are not born with a hammer-shaped head, their head is more rounded than their parent's. Pups face some predators but as they grow to their full size, other animals tend to leave them alone. After females give birth, they leave their young to fend for themselves. Then after the pups are born, they swim in a school together until they are old enough to separate and hunt alone.

African Congo Peacock
Nyale Keita

A peahen and her peachicks wonder off into big area with busy streets lots of people and houses. Soon three kids woosh by the family of peacocks. Two of the kids run on the sidewalks while the other child goes into the bushes. The family of peacocks follows the child who goes into the bushes. Once the child realizes she is being chased by five peacocks, she tries to out run it, but peacocks run 10-20 miles per hour. The average 11 year old girl only runs 5-9 miles per hour. So, those peacocks stayed right behind her. The peacocks continue to chase the little girl. The girl comes to an unexpected stop to open the gate with a huge yard full of things a peacock would love to have. All that was in the gate was all the things that a peacock would love like, grains, insects, butterflies, leaves, figs, and other things.

In the little girl's front yard there happened to be a mirror. The mother peahen had puffed up her feathers and began to hit the mirror until the little girl had noticed that when they see their reflection puts them into fighting mode. So the little girl runs and grabs the mirror away from the peahen. People start to stare as the male peacock fights another male peacock who followed the peahen and her peachicks. When most people see a peacock, they think of their beautiful feathers. Well, both of the peacocks had a green metallic neck and brown, green, black, and blue feathers puffed up. The peacocks had started to fight then pretty soon gave up.

Months had past and the peacocks had continued to live with the little girl. The little girl continues to provide the food for the family of peacocks. It was the beginning of winter, and the girl starts to think that the peachicks wouldn't survive this extremely cold weather. So the girl puts the peachicks are put into a warm, big sized and cozy barn that looks like a little bit of everywhere around the globe. It was full of places that peacocks live in. In one area, it had looked like Central Africa, and India the other corner looked like the rainforest and southern Asia. All the peachicks scattered around the barn. The little girl noticed that the mother wasn't in the barn with her peachicks. The little girl had heard cawing that sounded like a peacock. As the door is opened for the peahen, the peahen runs looking like she is full of enjoyment. Winter was almost over the peachicks were old enough to go hunt by themselves. So the little girl lets the peahen and her peachicks free in the wild.

The five African peacocks would occasionally being chased by four tigers in the wild. They would always puff up their feathers. When they puff up their feathers they have to be very careful because the tiger, one of their biggest predators, would give them a little peck behind. The peahen and her four little peachicks try and fly into a really high tree, but one doesn't make it up all the way. So the little peachick caws to its mother. As the little peachick caws for her mother, her mother comes and helps her little peachick. While the other peachicks are just sitting in the tree they see that a tiger was still in the area. The mother hears her little peachicks cawing for her. The mother soon sees that the tiger is focusing on her little peachick running up the tree. The mother scares the tiger, and she gets her little peachicks and begin walking.

Like they always did, they would leave the wild and give a small visit to the little girl. A peahen has a hard life in the wild.

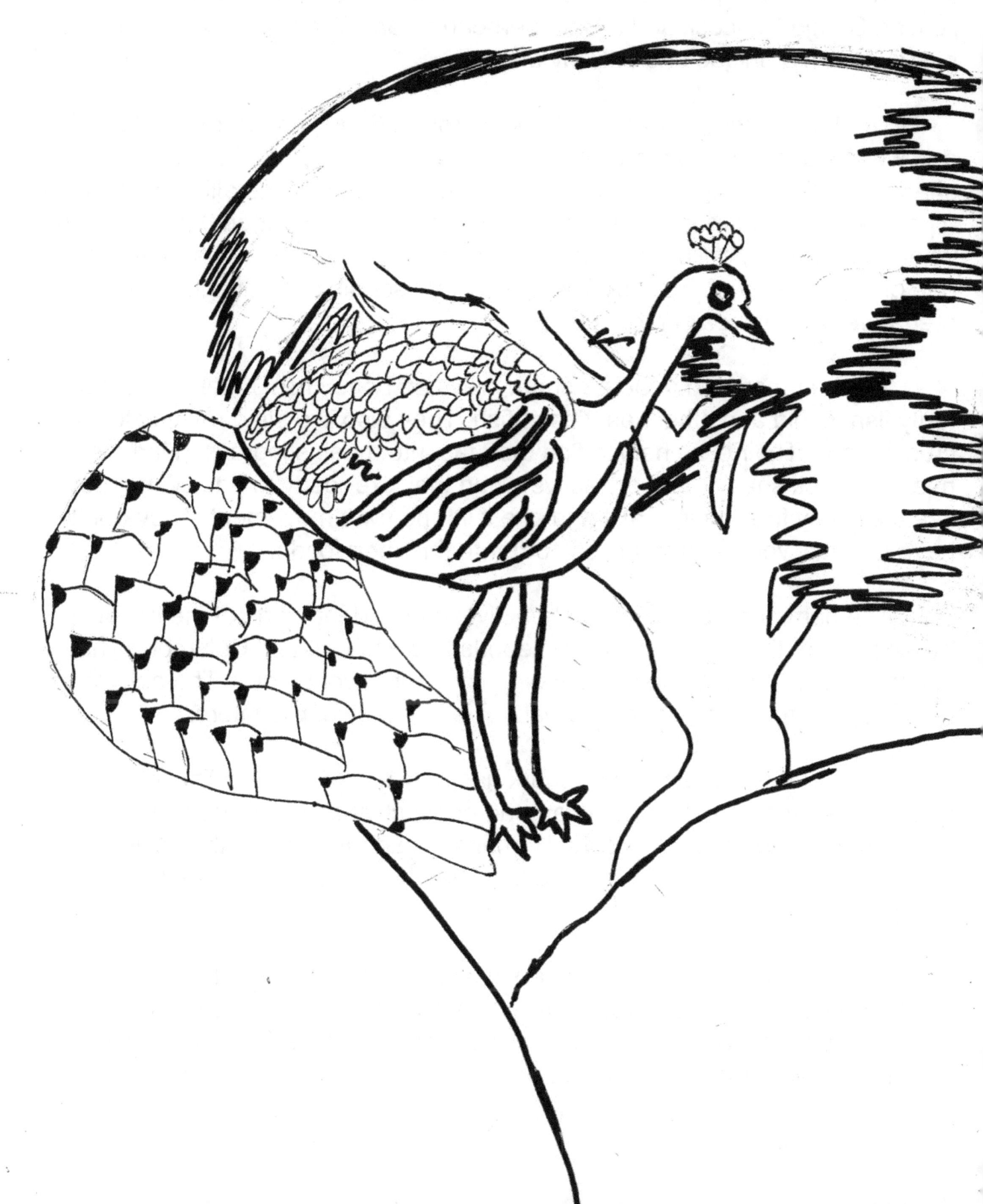

Hammerhead Peacock Shark
Nyale Keita and Miz Valverde

The hammerhead peacock shark is a animal that is a combination of an African Congo Peacock and a Hammerhead Shark. These two creatures combined to form a unique and unlikely animal.

The hammerhead peacock shark is around 5 feet long from head to tail and weighs about 150 pounds. It has the head and body of a hammerhead shark and the coloring of a peacock. They turn green and metallic blue just like a peacock. Hammerhead peacock sharks have a white spot on their belly. But when they are camouflaged their belly and their back turn into a beautiful and one of a kind array of peacock feathers. They use this camouflage to hide from predators and to sneak up on their prey.

The hammerhead peacock shark preys on boney fish, small crabs, butterfly fish, squid and other varieties of fish. They also like sting rays, octopi and crustaceans. Their eyes help to find their favorite meal: stingrays. Hammerhead peacock sharks are very active when they are ready to eat. When they see their prey they charge at it and bite it in one swift movement. They don't have to travel very far distances in order to find their food. .

The hammerhead peacock shark lives in the deep parts of the ocean by islands like the Cocos Island, Malepo Island, Hawaii, and Columbia. They occasionally leave the deep trenches to find their favorite food. The only other time they are forced to leave their habitat is because of predators.

The hammerhead peacock shark has only two predators. These predators are humans and other hammerhead peacock sharks. Humans hunt and eat hammerhead peacock sharks. Humans also hunt them for their skin and oils. Hammerhead sharks are also a threat to themselves because they are protective of their young.

The hammerhead peacock shark is a fascinating animal. Their appearance alone is very impressive, but it is their appearance that is making them a target for humans to hunt. These amazing creatures deserve to be protected from fishers and not used for their beauty.

Warthogs
Anthony Bahena

Warthogs are pig-like in appearance but their bodies are actually very different from a normal pig. Having neither fur nor fat, the warthog lacks both protection from the sun and insulation from cold. Warthogs have gray skin, thin legs, and long tails. Adults weigh between 130 and 250 . Their bodies are around 4 feet and males are larger than females. Warthogs have tusks that are 12 inches long. Warthogs have a bumps on their face that protect them during fights.

Warthogs are very hungry mammals. Warthogs eat mostly grass and other plants. They eat roots, bulbs, and thick underground stems of plants called tubers. A lot of warthogs use their strong sense of smell to sniff out food that they dig up from the ground with their tusks. Warthogs eat tree bark , fallen fruit and the flesh from dead animals called carrion. Unlike many other wild pigs, warthogs feed during the day and sleep in the abandoned burrows at night. Warthogs kneel on their front legs to search for food with their snouts. Instead of standing up and walking, they may shuffle from one grazing area to another in this kneeling position. During the wet season, they may eat earthworms and other small invertebrates.

Warthogs cannot be found in many places in the world. Warthogs can only be found in Central Africa. Warthogs live on savannas and in lightly forested areas. They always avoid living in rain forests, deserts and high mountains. Warthogs rarely build their own den, they often find another animal's abandoned den and claim it at their own. They live near water because they need it to stay cool when it's hot and they line their den with grass for warmth when it's cold.

Warthogs have some fierce predators. Warthog's predators are lions, leopards and hyenas. When chased by predators warthogs run to their dens and enter tail-first in order to protect the opening of their den with their tusks. Warthogs bend on their knees while digging for food, this makes them vulnerable to predators.Warthogs usually run away when predators come too near. They can run up to 30 miles per hour for a short period of time. Warthogs have excellent hearing and sense of smell which they use to smell or hear approaching predators. Warthogs are also vulnerable to human hunters.

Warthogs are social animals. They live in family groups called sounders. Sounders may include more than one female and young of all ages. Two male warthogs often fight for a female. The male who pushes the other warthog back wins and gets the female. Females give birth about 170 days after mating. The female warthogs stay in the dens with their newborn piglets for about one week. Piglets look like adult warthogs, but the bumps on their face are smaller and their manes are shorter. Piglets do not leave the dens without their mother.

Warthogs are often thought of as wild pigs. Even though they are related to pigs, they do not have a lot of the same features as regular pigs. Warthogs are fierce, fighting, speedy and alert creatures. They are smart animal that use all of these characteristics to survive.

Big Horn Sheep
Ehliyhan Eslava-Deanda

You are really lucky if you have seen a bighorn sheep because you are one of the few people that have seen one. It is really rare to see herds of bighorn sheep because they live in areas that not that many people go and because they don't like to be around humans. If you live in Canada though, you might be lucky enough to see some climbing in the mountains.

They live in mountains that have mello grass and rocky cliffs. Depending on the climate, they mostly live in the rivers and valleys. If it is winter, they live in the mountains, if it is summer, they live in the rivers and valleys. They sleep by curling up in a small pocket of grass or near dry water. They need rocky areas to escape from their predators. The area where they can live survive is in Canada.

The big horn sheep has to be really careful because it has a lot of predators. Their predators are cougars, Golden eagles, wolves, coyotes, bears, bobcats, and lynx. They can only escape by living in Rocky Mountains. They survive by hiding in the rocks and in the bridges.

The big horn sheep is an herbivore. They never eat meat. The bighorn sheep eats grass, shrubs, and forbs. In the winter, they eat woody plants, such as willow, and rabbit brush, plus a little bit of sage, blueberry and roses.

Their behavior is really grumpy especially if their predators are hunting or touching their babies. The bighorn sheep sometimes show-offs, especially the males sheeps. They do this to impress the female sheep because the females pick what sheeps are in what herd. Therefore, sometimes the male sheeps try to show the females they are responsible, by helping select sheeps for the herd.

The big horn sheep is a sensitive but strong animal. You can tell this by their features. The male horn is average 3ft 6in long and weighs about 40 lbs. The male height is 3-3 ½ft and the female height is about the same thing about the male, but just a little bit lighter. Their upper jaw is really hard so they can cut off the hard grass that they eat. Their lower jaw contains six biting teeth that are used against a pad. The rump of the female and of the male is used for signaling device and also for mating.

They migrate for a long time but it is really useful to them. The big horn sheep migration is really important but also really hard because in that time all the bighorn sheep predators are migrating and they might catch them. There migration is in the seasons of autumn and early winter. When the baby big horn sheep migrates, it has to be up to 1 - 4 years old. They migrate from the south coast to the northern mountains. While, they are migrating, they are mating for two months. They migrate because of the climate.

They are really special animals. They have been around for a long time. They live up to 20 years old and the oldest they could be is 25 years old. In 1948 only five groups of the big horn sheep remanded because during that time a disease was going around. In 1979 and in 1988, threes canyons fell down that also threatened their species. Since they didn't have a habitat their predators ate them. In 1999, ESA, helped them survive and saved their species.

The big horn sheep is really special compared to other animals in the sheep family. They are sometimes wild, but they are also sensitive. They can be really smart especially if you are teaching them something. They are really heavy and hard to take down especially if they are male because they have really big horns. These are just some of the reasons why I love big horn sheeps.

Hogzilla Sheep

Ehliyhan Eslava-Deanda and Anthony Bahenaio

If you have seen a hogzilla sheep living in the mountains you can consider yourself lucky because it is a rare and unique creature. It has the features of both a big horn sheep and a warthog.

The most striking feature of the hogzilla sheep is its two sets of horns. It has the horns of a big horn sheep coming out of his head and the tusks of a strong warthog coming out of its mouth. The horns are about 3.5 feet and the tusks are about 12 inches long. Adult hogzilla sheep weigh between 300 and 400 pounds. Their bodies are around 5 feet in length and males are larger than females. The hogzilla sheep has a little bit of fur that it uses to stay warm and that the males use to impress females during mating season. It has the strong and powerful body and feet of a big horn sheep that it uses to climb steep mountains. It has the tail and the snout of a warthog. They use this snout to sniff out their food.

Hogzilla sheep sniff out grass and other plants to eat with their strong sense of smell. They also use this sense of smell to find fruit and plants that are not affected by other animals. They find their food in the grassy mountains of the Savannahs. Hogzilla sheep eat woody plants such as willow and rabbit brush, plus a little bit of sage, blueberry and roses. During the wet season, hogzilla sheep may eat earthworms and other small invertebrates (an animal that doesn't have a backbone). When they are searching for food, they bend down on their front knees to make it easier to sniff it out. After it rains, they dig to look for worms because it is easier for them because the worms come out after the rain. They eat worms by crushing them with their knees then bending down and eating them. When they are focused eating, they easier targets for predators.

Hogzilla sheep predators include lions, leopards, wolves, coyotes, cougars, golden eagles, bears, bobcats, lynx and hyenas. To protect themselves from these predators they run away. Hogzilla sheep are really good at navigating the rocky mountains. They do this by having light feet so they won't get caught. When the hogzilla sheep is getting chased by predators they run to their dens and enter tail-first in order to protect the opening of their den with their tusks and horns. If a hogzilla sheep gets caught by a predator, they use their tusks and horns to fight back. They rarely get caught by a predators though, because they are so skilled at climbing and running through the mountainous regions they call home.

Hogzilla sheep can most often be found in the mountains near central Africa. They live near the mello grass which they eat to survive. They live near water because they need to stay cool and fresh. They only are in families when they are hunting. They don't communicate because this makes the other predators locate where they are. They make abandoned burrows of other animals their home. They use these burrows to keep warm and safe from predators.

Hogzilla sheep are a unique combination of warthogs and big horn sheep. They are rare creatures that use their strength and tough adaptations to survive.

Bottlenose Dolphin
Giselle Moran

I am a bottlenose dolphin some of you might have seen my relatives at Seaworld. If you were to try to find me it would kind of be hard because I live around the pacific Ocean out in the deep water. I also live around the shore of Hawaiian tropical Islands.

You can spot me because of my unique shape. I have six main things on my body those are my tail, blubber, flippers, eyes, blowhole, and my fin. My tail is used for swimming, it goes up and down not side to side. My blubber keeps the cold out and my flippers are for talking to other dolphins like you humans talk to you friends and family. I have two eyes just like you and a beak to eat. My beak is smaller than a regular dolphin, it is because I am a bottlenose dolphin. I have a blowhole that is on the top of my head it is where I breath from and last is my top fin that prevents me from rolling over.

I like to eat fish, squid, crustaceans, animal plankton and plant plankton. Usually the amount of fish that I eat depends on the kind of fish I hunt.While mackerel or herring have a lot of fat, squid have not much fat, therefore, to get the energy required for their activities, dolphins have to eat more squid than mackerel.I will eat between 10 Kg to 25 Kg of fish every day.I bet you humans eat fish tacos just like I eat fish.A cool fact about us is that sometimes we catch larger fish by hitting them with their tails to stun them.

If you were to try and hear us communicate it would be kind of hard because we have many different ways that we talk to each other.One way we talk to each other is by hitting are bodies on the water.We also can read hand signals.Some of the sounds that we make are buzzes, squawks and a pop.We also can make sounds to pass on information.

Snowy Owls
Makari Saucier

Snowy owls are diurnal, which means that unlike other owls they like to be active and hunt during the day and night. They often hang out at airports. They do this because of how wide the airport is.

The snowy owl is all white just like its habitat, the Arctic. They are attracted to large open areas, which is why they like the airport. But they mainly live in open, treeless areas called the tundra. They show up in the winter to hunt in fields or dunes. They spend summers far north of the Arctic Circle hunting lemmings, ptarmigan, and other prey in 24-hour daylight. Snowy owls like to be on the ground. They perch on the ground or on short posts.

Snowy owls are among the largest North American owl species. They weigh 40-70 ounces and their wingspan ranges from 49-59 inches. Males are smaller than females. They have big yellow eyes, a black beak and white feathers. Snowy owls are highly sought after by birdwatchers because of their unique appearance.

The snowy owl is a skilled and excellent hunter. To capture its meals, the owl relies on its advance sense of hearing. They hunt for small rodents, rabbits, birds, mammals, and fish. An adult owl may eat around three to five lemmings each day which is 1,600 per year. Snowy owls like to hunt in the night. Once they spot their prey, they approach it from the air, and snatch it up using the large, sharp talons, or claws, on their feet. Sometimes, if there is not enough prey around to feed baby owls, the adult pair won't lay any eggs at all until the supply of food improves. When snowy owls are hunting for smaller birds they go down and catch them while flying. Unlike most other owl species, snowy owls also hunt in the daytime. Snowy owls are highly nomadic and their movements are tied to the abundance of their primary prey species and lemmings.

They are known to aggressively defend their nests and will attack those that disturb their nets. They have very good eyesight but they can't see their prey when it underneath snow or thick layer of plants. The snowy owls predators include arctic foxes, corvids and jaegers.

Snowy owls are unique and popular among birdwatchers. Their beautiful white feathers, uncommon hunting habits and their ability to defend their young make snowy owls a rare creature.

Bottlenose Owl
Makari Saucier & Giselle Moran

The Bottlenose owl is a very interesting animal. It is a bird that can fly and swim. This is a unique feature because most birds cannot swim for long periods of time, but the bottlenose owl can.

The bottlenose owl is able to swim underwater and fly because of its unique appearance. They have the body of an owl and the tail of a dolphin. They have a bottlenose nose and flippers that make it possible for them to fly and swim. The bottlenose owl has a lightweight frame, which makes it easier to fly. The Bottlenose owl has both slippery skin and feathers so that it can seamlessly transition from air to water when it is diving for prey.

The bottlenose owl eats fish, birds, squid, and small rodents. Usually, the amount of fish depends on the kind of fish they hunt. They eat different kind of fish like, salmon, herring, cod, and mackerel. They use their bottlenose nose to dive in to the water and get the fish. The bottlenose owl will eat between 10 Kg to 25 Kg of fish every day. They mainly prey on fish and birds because there aren't many rodents in their habitat.

The Bottlenose owl lives in the Arctic. They like cold temperatures and cold water. Even though the bottlenose owl spends most of its time in the water, it makes a home for itself on land. This home is a burrow dug out of snow. The bottlenose owl will live in this burrow until it spots any predators that could possibly be a threat to their home.

Bottlenose owl predators include arctic foxes, wolves, orcas and people. They protect themselves from these predators by staying in groups of 10 to 12. These groups are called plocks. These plocks help keep the bottlenose owls protected because when they are feeding, they can all look out for predators.

Not only is the bottlenose owl a unique type of bird, it is a unique animal in general. Its extreme habitat and rare appearance make it interesting and special.

Elf Owls
Delanie Gomez

Elf owls are very interesting creatures because they are the lightest owls in the world. Although they can't see all the way around their bodies, they can see in the dark. These creatures soar through the sky while holding their light weight body.

Elf owls can be found in many places in the world. They are mostly found in riparian habitats (places where there is water) or desert environments where they can get water from things like the Saguaro Cactus. For example, one place they can be found is the Sonoran desert. Elf owls can live in mountain slopes, canyons, ravines, and plateaus. They live in many different places including Baja California, Southern Arizona, Central Mexico, Pacific Sinaloa, Socorro Island and southwest U.S.A.

Elf owls don't have a lot of predators. They threatened by other owls, snakes, coyotes, bobcats, and ringtails. It's really difficult for predators to access elf owls because they nest up high in Saguaro Cactus. They live in the Sonoran desert because they eat a lot of cactus and there are a lot of cactus there. Elf owls diet includes small lizards, locust, grasshopper, beetles, moths, crickets, caterpillars, mantids, centipedes, cicadas, and fly larvae. By catching their prey first, they fly around and listen to the heart beats of mice then if they hear anything they plunge after the mice. Last, it sinks its talons into the the prey.

Elf owls have different features from other owls and do not have ear tufts or feathers at the top of their rounded heads like many other owls do. Elf owls eyes are pale yellow highlighted by a white circle. Their horns are gray with colored tips. Their feathers are grayish brown. They have thin bright white eyebrows on the top of their eyes. Also, Elf owls are the smallest owls in the Sonoran desert. Elf owls are five inches tall and their wingspan is 9 inches. Their weight varies between 1-1.5 ounces. In the wild, elf owls live 3-6 years, but in captivity elf owls can live 6-10 years.

The elf owl, despite being the smallest still has a lot to offer. Their rarity and beauty makes them an interesting sight to see and should be kept alive.

Snow Leopard
Makaia Saucier

Snow leopards can only be found in certain areas in the world. In the wild, their average lifespan is 15 to 18 years. Like most big cats, leopards are loners because they spend most of the time by themselves. However, they can communicate with each other by scent-marking. Leopards are fast and fierce animals that are at the top of the food chain.

Some of the snow leopards eat blue sheep from Himalaya. Snow leopards also eat birds & bears. They hunt eat animals such as Rodents, monkeys, wild sheep, and goats. They also eat marmots, wild boar, bobak, tahr, mice, deer and game birds.

Snow leopards are found in the central Asia. They can be found on the highest mountain named Mt. Everest. They also live in the Alpine zone in the summer and they come down the Subalpine in the winter. They sleep in cold places such as dens, trees, and mountains. Their fur is very thick to help them preserve heat as they live in extremely cold. In the summer, they prefer to stay in the meadows and rocky areas.

Snow leopards have some interesting features. When snow leopards are in danger they climb trees. Their coat provides protection by camouflaging it from their predators. When leopards are not on the prowl they can sleep up to 20 hours a day. At an early age, cubs learn the hunting skills of surviving. Snow leopards can use their long tail to keep its nose warm. They can move so silently that sensitive ears of an impala cannot detect it. Mother leopards can communicate by puffing & sniffing.

Snow leopards have a lot of threats. Some of their threats people use from their skin are their bones, other body parts, clothing, rugs, and traditional medicine. Snow leopards are facing a distinct threat from global warming. Snow leopards predators are similar animals such as other large cats like tigers and jaguar. One of their other threats come from humans. Hyenas can kill younger leopards and drive them away. Baboons can gather up a whole group to attack and possibly kill a single adult leopard.

Snow leopards are found in cold climates. They have a lot of skills that help them survive in the wild. Because of their beauty they were sometimes hunted by humans. It is important to keep these animals safe!

The Snow Elf
Makaia Saucier and Delanie Gomez

The snow elf isn't Santa's helper, it's a tiny exotic creature that is part elf owl and part snow leopard. It can mostly be found in the snowy area of Asia. It has powerful vision and an extreme sense of smell, making it a fierce predator. The snow elf is extremely rare and a unique looking creature..

Snow elves are beautiful winged animals. They have dark grey spots and a white underbelly to blend into their surroundings. Snow elves have feather tufts at the tips of their ears on their rounded heads to hear better. They tend to be 30 cm to 50cm long with a tail of the same length. A snow elf will weigh between 60-70 ounces. Snow elves have thin bright white eyebrows at the top of their eyes. They have talons to help them catch prey and perch on tree branches. They do not have a beak like other birds, but a nose and mouth with sharp teeth to help them sniff out and capture their prey.

Snow elves are carnivores. They eat rare food like rodents from Himalaya. Sometimes they also eat birds and hares. The snow elve's diet includes small lizards, locust, grasshoppers, and other insects as well. A snow elf hunts by listening and smelling for prey. When they hear their prey make any noise, they swoop down to it sink their claws inside of it and devour it. They are excellent hunters, but have many predators themselves.

Snow elves are difficult for predators to catch because they nest high up in trees. They do have many predators including other owls, snakes, coyotes, bobcats, and ringtails. Although snow elves can't see all around their body they have excellent night vision, which they use to avoid predators at night. They also use their excellent sense of smell to avoid areas that are inhabited with their predators. Snow elve's main form of protection from their predators is their ability to fly. Snow elves soar through the sky holding their light weight body to get away from predators and to migrate to different locations.

Snow elves are found in the snowy mountains of Central Asia. They live in the Alpine zone in the summer and then comes down to the subalpine zone in the winter. They are found in places near a fresh source of water. Snow elves build their nest out of twigs, leaves and their own feathers. They build these nests high up in the forest trees. Unlike other birds the snow elf doesn't lay eggs. It gives birth to 1-2 babies per year that it nurtures high up in the nest, away from any predators..

The snow elf's babies may be cute and cuddly creatures, but the adult snow elf is actually a fierce and strong hunter. Even though it is a very tiny bird, it's cat-like features help it hunt like a it is a large predator. Its beauty, power and appearance contribute to the uniqueness of this rare creature.

Striped Dolphiin
Cameron Siry

Dolphins are among the most intelligent creatures in the animal kingdom. Striped Dolphins are a unique breed of dolphin. They use their intelligence to get away from predators, communicate with each other, and catch their food. They are one of the most fascinating mammals in the ocean.

They are social creatures that live in pods of 25 to 100. They communicate by making clicks, whistles, and using body language. The clicks are used for echolocation, which means sending out a sound and determining distance of objects or prey based on the time it takes the echo to return.

The striped dolphins they live in different regions where all their food is found. Their diet is bony fish like tuna and cod. They also eat crustaceans (crab, lobster, crayfish, or shrimp). They like eating celephods, such as squid or octopus and eating krill. They feed where there is a group of fish. They teach their babies how to get their own food by taking them out on a journey and showing them how to hunt baby fish and how to blend in when predators come. Even though the striped dolphin doesn't have lots of food sources, they can still live off of what they find.

The places where striped dolphins can be found are in the Mediterranean Sea, the Pacific Ocean, the Atlantic Ocean, the Indian Ocean, the Caribbean Sea, and in the Northern Gulf of Mexico. However, they are most likely to be seen in the Atlantic and Pacific Oceans. They are usually near the shore on the shallow part at the beach where people can see them and enjoy them splashing and communicating!

Striped dolphins are easy to identify. Their body length is between 2.2 and 2.4 meters. They have long narrow blue flippers. A blue-black stripe goes along the length of their body. They have long black stripes on their tail, and they have stripes along their body that makes them look like there are waves on their body. They also have extra fins on their backs for swimming faster to get away from predators. Striped dolphins are often mistaken for sharks because of how they look.

The main predators of striped dolphins are humans. Humans kill dolphins everyday. Sailors sell dolphins in China for $100-$150 each! Sailors and tourist kill dolphins because they think it will get in the way of their fishing cast. Dolphins get accidentally killed by fishermen when they get caught in fishing nets. It is important to prevent this by doing dolphin safe things like buying "dolphin safe" tuna.

Axolotl
Eric Schroeder

The axolotl is a type of salamander whose species is slowly dying out. Its habitat is shrinking along with its population. This an endangered animals is an odd and often forgotten about creature.

The axolotl has pink, feathery gills on each side of its head. They are commonly found with black/brown skin but are also found albino and have very pale pink skin. They have short finned tails that are similar to a fish tail. Axolotls keep their same appearance from when they are babies and remain a giant four-legged tadpole as they grow up.

Axolotls are found in Mexico. They are currently only found in lake Xochimilco, but used to be found in Lake Chalco before it dried up. They also used to be found in swamps, lakes, marshes and temporary wetlands. Because their habitat is shrinking, so is their population.

Axolotls eat a wide variety of food. Its diet includes, mollusks, insect larvae, worms, small fish and crustaceans. The axolotl has a meat-based diet. They have very basic teeth which are used for gripping its prey instead of biting which causes them to swallow their food whole.

The axolotl is an endangered animal due to lake inhabitants like carp and tilapia. This is because those animals eat their eggs. Axolotls can regenerate body parts (grow back) and use visual and chemical cues to communicate. When they get old, they do not lose their external gills unlike frog tadpoles and produce self-sufficient babies.

The axolotl is an endangered but interesting amphibian. They live Mexico, can regenerate their body parts, and come in different colors. Hopefully they will remain in their habitat so that they do not become extinct.

Daxololphin
Eric Schroeder & Cameron Siry

All over the world there lives an odd animal. Its name is, the daxolophin or the daxophin. It was once endangered but is now extremely popular all over the world. Every year there are more and more daxololphins living in all parts of the world.

They used to only live in Lake Xochimilco but the species later expanded to oceans all over the world. Some of the oceans include, the Indian Ocean, the Atlantic Ocean and the Caribbean Ocean. They also live in, marshes, small lakes, wetlands, and fresh bodies of water. They live in the sunlight zone of the ocean because it needs to go up for air and there are a lot of predatory fish there. This can also be dangerous though, because humans mainly fish in the sunlight zone. The daxololphin looks very different from the other fish in their habitat.

The daxololphin has a very strange appearance. They have feathery pink gills on the side of its head that helps it breath underwater. They keep their appearance from childhood and do not lose their gills like a frog does. They have a tail which is similar to a whale's tail and have long, blue-black stripes that run along its back and fins. They have long and narrow fins. The daxolophin is commonly found from 8-12 inches long. They are sea mammals and have a blowhole on their backs but need gills for a reason.

They use their gills for deep dive so they can breath for a longer time. Sharks rarely feed on them and large fish eat their young. Humans are also a threat to the daxololphin. Fisherman catch them with large nets and sell them for a lot of money. They produce self sufficient babies, and they can regenerate their body parts. They can use clicks and whistles to communicate but can also communicate through visual and chemical cues. One reason they communicate is to alert each other when food is nearby.

They eat animals that can be found in the ocean including small crustaceans such as, crayfish and small crabs. It also eat things like, small fish, insect larvae and krill. Their diet also includes, mollusks, cephalopods, and worms. They teach their young how to hunt at a very young age. They catch their prey by slapping their tails against the water to stun their prey. They hunt their prey in the water.

In conclusion, the daxololphin is a very interesting sea creature with odd characteristics. Some of these characteristics include being a mammal with gills and it looks very different from other fish. The daxololphin is a very strange animal.

The Gray Wolf
Jesse Stephens

I am a gray wolf and this is my story. My head peered around the leaves and I ran to the den after a hunt with an older wolf. The older wolf was one of the two alpha wolves. He was my loyal and trusting father that helped raise me as his own. A young wolf like me was raised by the pack until I was about ten months old because baby wolves are born blind and defenseless. It was my responsibility to help the alpha wolves to help take care of the pack. Later that day, the pack and I came to back to the den after a hunt when we realized that one of the pups got lost. I was the beta wolf which meant I was the second strongest member of the pack so I ran into the forest to find it.

I had to find him fast before the sun went down and got to dark. The reason I was the beta wolf was because with my thick gray fur helps me withstand harsher and colder weather. Also because of my strong ears that allow me to hear better. The alpha wolves made me beta wolf for all of these strong features.

Us gray wolves have a great sense of smell so I used that ability to track down my pack member. Finding him was hard because my pack lived in the cold part of Alaska and there were a lot of fallen trees. Pretty soon, I found him near a big tree. The tree that divided the pack territory with another wolf pack. Luckily, I was able to find him fast because if I didn't he could have crossed into another wolve's territory and they could have started a conflict with the packs. Members of the pack will risk their lives to save other wolve's, so I quickly ran to him and stopped him from crossing over. He went on his back and put his paws in the air because his knew that I was a more higher rank wolf than him.

Soon, we came back to the den. It was almost sunset so the rest of the pack was gone because the pack is most active during sunrise and sunset. But before I went in the den, I used my howls to tell the other wolves that I found the lost pup. I heard their howls as I stood on the top of the den gazing into the sunset. Then, I went in the den with the other pups for the evening. When the rest of the pack came back they went to different spots of the den and slept. When we woke up the alpha wolves and some more of the older wolves were training and teaching pups new things that they can use in the wild. We had strange way of teaching the pups, but they learned.

I went out into the wild and found the food I was hunting for with my dad a few days ago. We left it there is because some food was too big to take home, but I brought as much as I could. It was mostly meat but the pack probably wouldn't mind. As I dragged the food, the wolf pup that got lost came and helped me push it up the hill. He did it because wolf packs always use teamwork.

The next day, the alpha wolves made an announcement that the pack would not be living in the den and that because of the previous event of a pack member going missing, we had to go somewhere more secure and safer. Soon we saw another pack and my dad went to ask them where they were going. Soon we came up a hill and saw a place called Yellowstone. The pack came to a new den, but as soon as we did, I saw something strange. It something that stood on two legs and only had hair on its head and it was carrying a metal object that was capable of making bright flashes and clicking sounds. Then more of them came and used the same object. I simply ignored it and walked into my new home.

Komodo Dragon
Jonathan Quiroz

Komodo dragons are fierce and territorial reptiles. Their black, green, grey, and yellow scales make them easily recognizable. However, they might be hard to spot because of their unique ability to camouflage. They are the largest reptiles in the world.

The komodo dragon has many different features that it uses for survival. One feature is, it uses its tongue to sample the air like most reptiles. Another feature is, to warn other animals if predators are near the area, by arching their neck and puffing their throats out. They also have shark like teeth that they use for hunting. While it hunts, it uses its poisonous spit to harm its prey.

The komodo dragon is alone from birth. They do not live in groups. They aren't social animals. When they hatch from an egg, they are all alone in a bird's nest. As a baby, they're only 12 inches as long as ruler. When they're adults they are up to 10 feet long. They can live up to about 50 years old.

Komodo dragons have a big appetite and eat a lot of different types of food. They eat wild pig, buffalo, snakes, and fish that wash on shore. They will eat anything sizeable and healthy. When it bites its prey, the bacteria in its mouth leaves a deadly infection inside the animal. They kill from the back and bite the legs of its prey to slow it down.

Komodo dragons live in exoctic places in the south of Asia. They live in grasslands near forests and beaches because they need water and trees to survive. They live on the coast side of Indonesia. When they hatch, they climb the nearest tree for protection from other komodo dragons, which are there only natural predators.

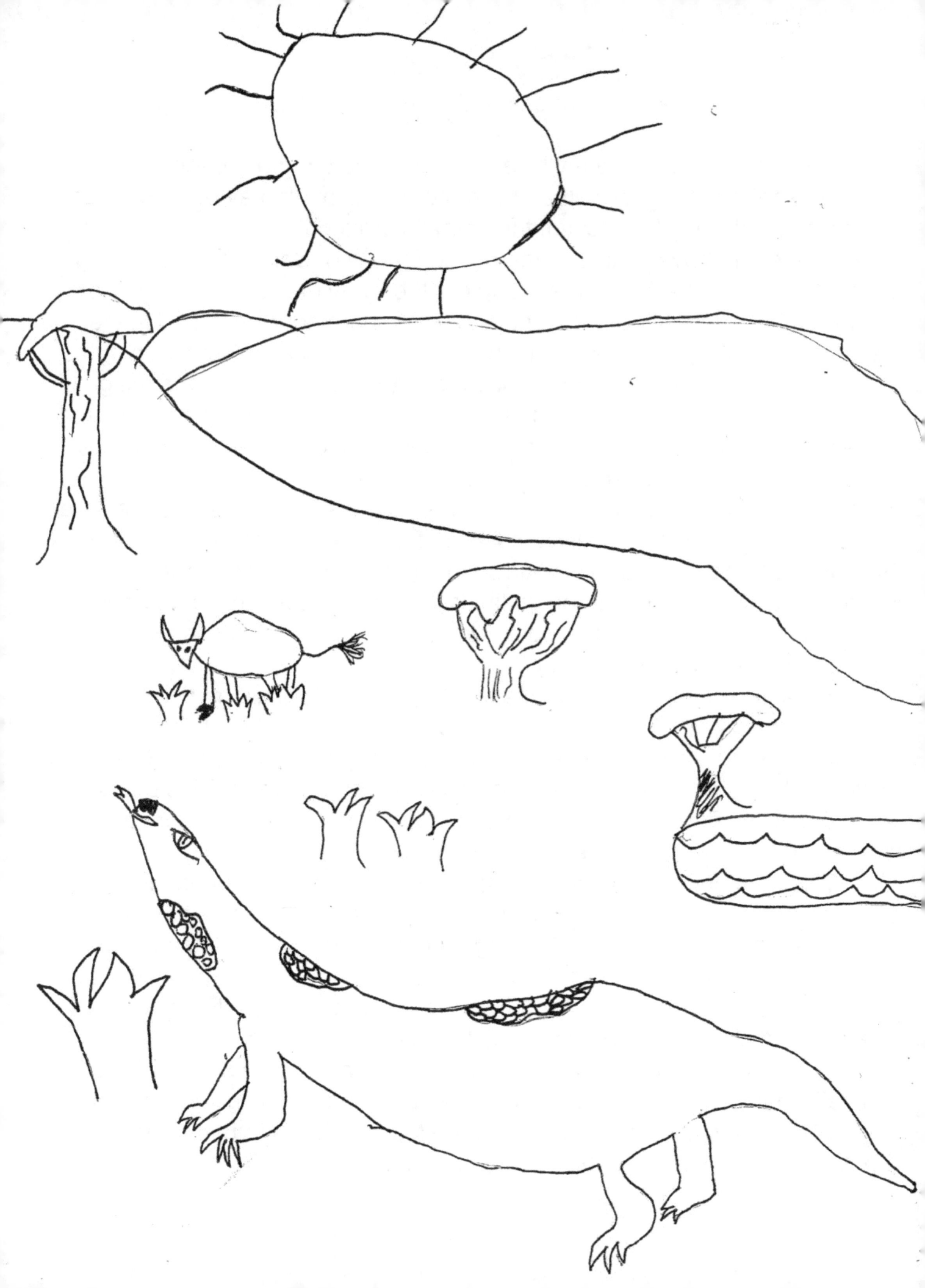

Gray Dragon
Jesse Stephens and Jonathan Quiroz

This story is about the mythical animal called the gray dragon which is a combination of a gray wolf and a komodo dragon. It may be an vicious mammal, but it is kind to most of the animals around it.

It lives and travels through the humid forests of China. The gray dragon often lives and travels with a group of other gray dragons and they wander through the forest together. They are often born near trees and in the trees because they can't defend themselves until they are about 10 months old. Sometimes when they travel, they move through big trees and tall grass to get to the cave or area where they live. They will use teamwork to help protect members of the group because the forest can have many dangers like falling trees and predators.

This creature has many unique features that it uses in the wild and they are all useful in its habitat. It uses its nose sniff its prey from far away. It uses its tall ears for better hearing of its surrounding. It has several different colors such as gray, yellow, and green on its scales and it uses these to camouflage in the forest. As a baby, a gray dragon is twelve inches, so they must be protected while they're young.

Because the gray dragon lives in a large deep forest it must adapt to its surroundings. It uses its four shark like fangs to defend itself from dangers in the forest and uses its sharp claws to keep balanced and move better on the forest ground. It uses its thick fur to keep warm. This unique animal has two pairs of lungs, one pair is for helping it breathe more and the other pair is connected to the heart to survive.

This creature travels through the forest, but to survive it must find food. The gray dragon often eats a strange animal called a piggbit which looks like a rabbit combined with a pig. The gray dragon keeps hunting and finding food in the forest like fish by a lake or the gray dragon may chase a piggbit running in the forest in hope of catching a meal for its packs. This creature's diet is strange but to them, it is quite delicious.

In conclusion, this is a very fierce, powerful and brave animal. This animal's many features are good for defending itself in the wild. Its adaptations help it survive in its habitat. It can be found in deep humid forests of China, traveling with its group of gray dragons.

Frigate Bird
Marco Galvez

The frigate bird is a large and very uncommon species of seabird. They are known for stealing food, flying for long periods of time and the distinctive red pouch found on the male's throat. This is just a few of the many facts about the frigate bird.

The frigate bird is very large and unique looking creature. They have the largest body to wing ratio of any bird in the world. Some frigate birds can even have a wingspan of 7 feet from tip to tip, which is the size of some boas constrictors. Because of their large wings they can stay up in the air for up to a week. Frigate birds are black with an iridescent purple sheen. Female's are slightly larger than males and their feathers are a bit lighter in color. The male frigate bird has a red pouch on their throat which they inflate for mating season. The frigate bird is thought to be related to pelicans. Even though they are a large species of seabird, they can't swim, and they can barely walk.

Out of the five different species of frigatebirds most live near the Pacific Ocean. Most of them like to inhabit tropical islands. Some can even be found on the Galapagos Islands. Frigate birds live near the ocean because the majority of the food they eat is fish. Frigate birds often share habitats with colonies of other seabird species. The nest they make consists of a loose assemblage of sticks, usually at the top of a tree or bush but sometimes on the ground. During mating season a frigate will gather sticks and choose a nest site and wait for a mate. Once the egg is hatched the male and female will watch the chick for the first few months until the male leaves and the female is left to watch the baby bird alone.

Frigate birds have interesting eating habits. Since about 10 percent or less of the food they eat is stolen from other birds they're known as pirates of the sea. They hunt by plucking fish out of the water gracefully with their hooked beak without even getting their feathers wet. The fish that the frigate bird hunts from the water can not be too heavy because these birds can not swim. During mating season a frigate bird may eat the chicks of other birds. They will also eat jellyfish and steal baby sea turtles.

Since frigate birds spend most of their life in the sky they have very few predators. Humans are their main predator. Recently frigate bird populations have been declining because humans have been destroying their habitats in order to build houses and resorts, causing a disturbance in their colonies. Even though adult frigate birds don't have many predators, their young and their eggs can be vulnerable to rats, lynx's and stoats.

The frigate bird is a extremely interesting animal. Humans thoughtless destruction of the frigate birds' habitats are causing this species to die out. Humans need to be more careful of their effect on frigate birds and their habitat.

Ring-Tailed Lemur
Sam Kahn

On earth, if we zoom in near the bottom left of Africa, there is the island called Madagascar. In the tree tops of its lush forests, is a ring-tailed lemur about the size of a house cat. Sweeping through the trees, it is a beautiful creature well known for its black and white ringed tail. Jumping from branch to branch, he finally comes to rest on a nice big tree trunk low to the ground. He then begins to lie down for the night. Just then, a foosa quietly walks among the trees, ready to pounce on the unsuspecting lemur. Luckily, the lemur was about to get up to get a drink from a nearby river. S o right when the foosa pounces, the lemur hears and quickly runs straight up a nearby tree where it is safe.

When morning comes, our little friend goes out to look for food from the favored tamarind tree. The tree holds within its branches the lemurs favorite snack, tamarind pods. As he approaches the tree, he sees other lemurs have already eaten all the pods. But again, luckily, on the other side of that tree, is four more of the tamarind trees. So he eats his dinner and finds a spot to sleep where he will sleep every night until the supply of pods run out. Then he will at that point he will have to get pods off of a different tree. But he will not starve while looking for new trees, because they also eat berries, plants, and sometimes small animals.

Just then, lightning strikes the ground. It's one of those dangerous tropical storms. The thunder was loud and lightning was everywhere. Our lemur has spotted a hole in a tree and he leaps into it! When the storm subsides, our furry friend comes out of the hole to see that the world is covered with greenery. All that rain gave the plants so much water that they might have had too much and that made the plants super green. Later that day, the lemur walks among the plants of the rocky hill he is on. In this place there is little food. But lemurs also eat small animals like mice, and this is a good spot to find them. Suddenly he hears a small noise in the grass. It's a mouse! the lemur stalks it very carefully with swift and light movements and then, BAM! It is caught and he now has a delicious meal for that day.

The next day the temperature goes super high and is raised to 104 degrees and so the ring-tailed lemur takes this as a time to sunbathe. This is one of their favorite activities as they need a way to get warm after a cold night. When there is temperature like this, it is an excellent time to. As he begins to sunbathe, he notices other lemurs around him doing the same thing, so this is probably one of the best days to sunbathe.

These things we saw the lemur do, were only a fraction of what really went on in his life. He lived to be 19, which is the most amount of time that a lemur can live. But this lemur was not a real lemur, he was just in imaginative character. But what he did, was mostly what real lemurs do.

Ring-tailed Bird

Sam Kahn and Marco Galvez

The ring-tailed bird is a strange creature. Not only is it one of the only birds to have a long tail, it is one of the only birds to have both feathers and fur. This fowl is in danger of becoming extinct. It is a rare creature that has features unlike any other birds.

The ring-tailed bird's long furry tail allows it to stay in the air for a long time. The long tail also allows it to hang upside down from a branch for long periods of time, as well as, helping with flight control, speed, and agility. It has large wings to help it fly when below the treetops. The large feathers it has on its wings help with control too and can even make the bird go from 40 mph to 60 mph in seconds. The color of its body is mainly a silver-grey color while the tail is black and with grey rings. Its six feet in length and two feet of that is made up of the tail therefore, it can't fit into small spaces. The bird also behaves oddly as well.

One of its behaviors is that when its a sunny and hot day, it likes to sit on a rock and clean its feathers. Also, when the weather is cold and rainy, it can't fly so it finds a hole or a cave and sits in it until the weather lets up. When mating, it makes a nest, and calls out with a loud screech to find a mate. When he does find a mate, if there is more than one then the females will fight to determine who will be the males mate. It is rare for two to come though because of their habitat.

The reason they have such a loud call is because they live in the Amazon rainforest and are a rare bird to see. Often in this area it is cloudy because of it being a tropical rainforest. Most of the time it's cold, but on rare occasions the weather turns hot and all the animals are happy. Where it makes its nest depends on how high the branches are on the tree. If its too low, then predators like a snake might get to to them. If its too high, then their babies could fall off and die. The ring-tailed bird doesn't have many predators. Their main predators are humans. Human's pollution is starting to have an effect on the ring-tailed bird's habitat. Their habitat is very large, and it can supply all the food they need.

One of their favorite food is the fisharind pods, a blue fruit that comes from a tree thats called the fisharind tree. To reach this fruit, it flies at one of the pods and rips it off the branch, where it then eats it. Other things it eats are small animals like mice and other things like that. It is also a scavenger, when it sees a large animal take down its prey, after it leaves it goes in and scavenges what is left. When this animal is full, it lies down to rest for the night.

The ring-tailed bird is an endangered animal that should be protected and preserved. The pollution by humans of their habitat has become an issue for the survival of the ring-tailed bird. This large majestic winged species could be threatened into extinction if nothing is done about it. icup

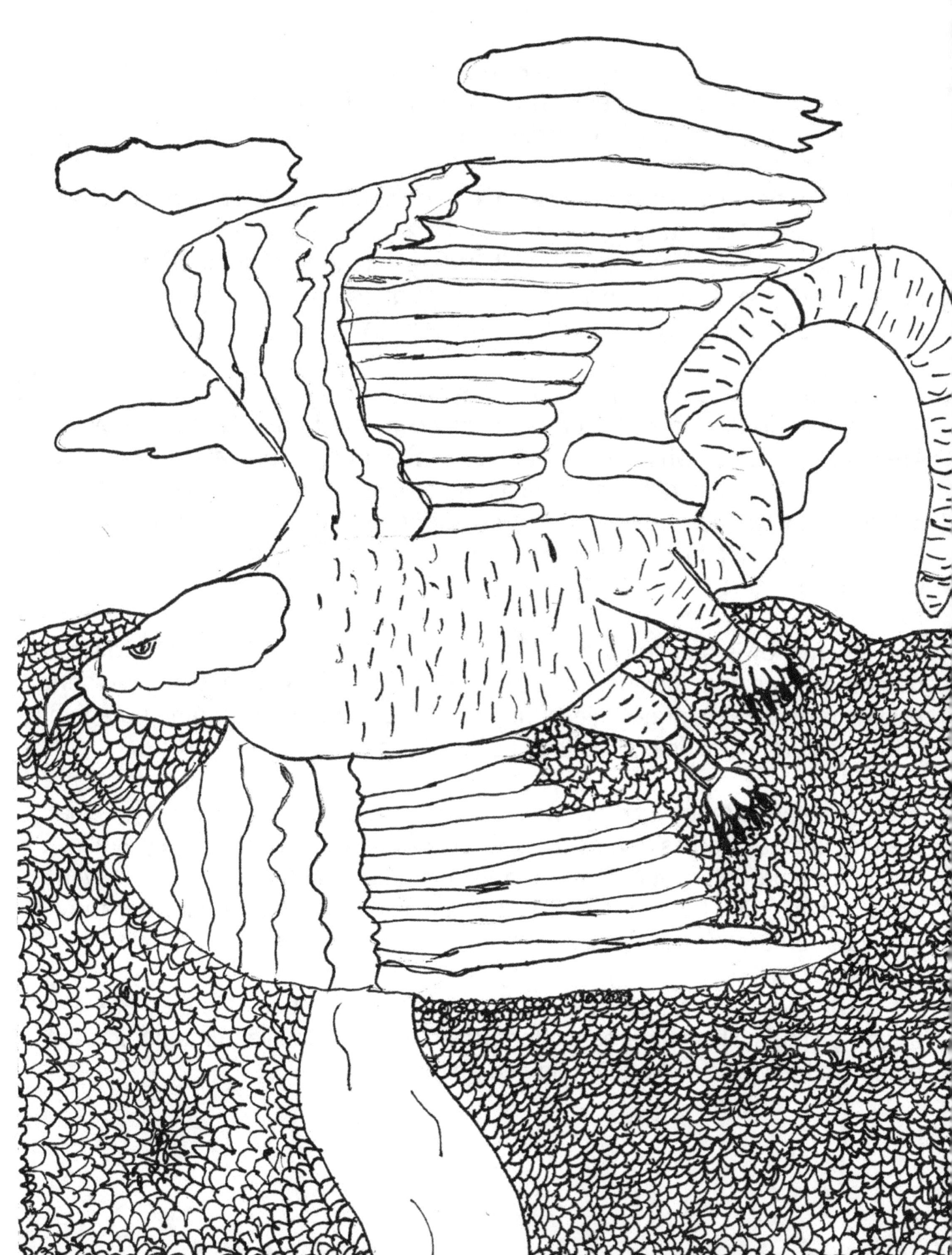

Cactus Ferruginous Pygmy Owl
Miranda James

    Cactus ferruginous pygmy owls are tiny fascinating creatures with many wonderful qualities about them. They have a lot of different animals they eat, places they live, and protection skills..

    Unlike their size, pygmy owls have very long names. They average at 6-7 inches tall. The males weigh about 2.2 oz. and females weigh an average of 2.6 oz. These fascinating creatures have longer tail feathers than other owls. They have reddish brown feathers with a cream colored belly. Pygmy owls have big yellow eyes that help them see their prey better.

    Small birds, lizards, and small mammals are included in its diet. Also included are frogs and earthworms. Pygmy owls hunt from dusk until dawn.. They are nocturnal which means that they hunt at night and sleep during day time. They have lots of skills for hunting, camouflaging, and defending themselves. They wait for their prey to come and then they go for the kill. Although, if these owls can't find food in their current location then they go search somewhere else.

    Pygmy owls can be found in a lot of different places. These little owls can be found in Arizona, Mexico, and Texas. They live in woodlands and desserts. Pygmy owls don't migrate unless they can't find food. If they come up against prey, they can use some skills they have for protection.

    They have many different ways to protect themselves. One way, is that they have small bodies to help them fit through the small holes they live in. Flying from predators very fast, camouflaging their feathers to blend in with surroundings, and holes in their head for powerful hearing are all ways they use their features for survival.

    The pygmy owls might be small, but don't fooled because they are very skilled animals. Their interesting features and ability to live in many different places in the world make them unique animals.

Red Fox
Yesenia Marenco

When you think of a red fox you might think of the song "What Does The Fox Say?" Red foxes don't dance, but the noises in the song do relate to their communication. They gekker to communicate, and yelp once or twice to aware family members of danger. Thats just one fascinating thing about a red fox though. There's more interesting things about a red such as their diet, their physical appearance, their social lives, and their habitat.

Many people can recognize a red fox by their unique features. Two of their most distinctive features are their long bushy tails and beautiful red-orange fur. Red foxes are small dog like mammals with sharp pointed faces and a slender bodies that allows them to be quick on their feet. They also have dark muzzles and ears, their tail tip, throat and other underparts are white. Males and females look very similar but the male is slightly larger. The males are can be called " dog, renard, or tods", while the female is called a "vixen". The jaw is more than half the length of its head. The head and body length ranges from 455 to 900 mm, the tail length ranges from 300 to 555 mm and their weight can be 3 to 14 kg. Red foxes are part of the family "canidae" which involves dogs, wolves, and jackals. Red foxes are rarely spotted alone.

Red foxes are social animals that live in small groups. Foxes typically live up to 3 years in the wild but can live up to 10 years in captivity. Females have about 4-6 cubs. The young juveniles stay with mother for 2-3 weeks. The cubs weigh 56-110 grams and their length is about 14.5 cm including the tail (7.5 cm). Males and females both nurse their cubs; however the mom does most of the nursing. The cubs open their eyes at 13-15 days old, their coat begins to change within 3 weeks, the cubs leave their den at 3-4 weeks. Red foxes always den within a few hundred yards of water, whether it is a stream or a pond, or a marshy area from after a heavy rain. Red foxes are careful to choose den sites that won't get flooded. They leave their den to hunt.

Red foxes go hunting at morning or dusk. They eat small mammals, birds, insects, and fruit. It also eats carrion such as road kill; scraps and waste. Foxes eat several pounds a day and bury the rest in the ground for later. They are omnivores so they mainly eat roots, rodents, plants, mice gerbils, ground squirrels, voles, woodchuck, deer mice, water flow and hamsters. Their predators are bears, coyotes, gray wolves, mountain lions, eagles and humans. Red foxes find this prey in their habitat.

Red foxes can be found throughout the northern hemisphere from the arctic circle to central america, as well as central Asia and northern Africa. This species has the largest distribution of any canid. Red foxes can also be found in Australia and falkland islands. These foxes utilize a wide range of habitats including forest, tundra, prairie, desert, mountains, farmlands and urban areas. They prefer mixed vegetation communities such as edge habitats, mixed scrub and woodland. They are found from sea level to 4500 meters elevation. The red fox builds a den for breeding and makes sure it has more than one opening in case of danger. When the fox isn't breeding, it sleeps in the open and keeps warm by wrapping its tail around it.

In conclusion, a red fox doesn't need dance moves to be a huge music hit, a red fox is a fascinating animal. It has interesting things about its life like it's habitat, communication, social life, and diet.

Pygmy Fowlx

Miranda James and Yesenia Marenco

    The Pygmy Fowlx, (also known as the fowl) is a very common animal that has great skills for surviving. But, that is just one interesting thing about this animal. It can be found in many locations and it eats many things. The Fowlx has to protect itself from many predators even humans because they will try to hunt it for the best tasting meat it carries on its body. This animal is unique because it looks very different from other animals.

    This creature has the tail and head of a red fox with an owl's cream colored body. Their weight ranges from three to five pounds. It usually has two black fox paws and two owl wings. Sometimes, if you're extremely lucky you can spot one with four legs. This means they are either deaf, blind, or both. For example, when a person is deaf another one of their senses might improve like their eye sight. They also have big yellow eyes for them to see their prey in the dark.

    The Pygmy Fowlx goes hunting from dusk to dawn. It is also an omnivore. This means that they eat roots and some small mammals. This creature can eat several pounds a day and bury the rest for later. It also eats earthworms and small birds. They kill their prey and find them in many places.

    The Fowlx can be found in many places like Arizona, Texas, deserts, and Australia. It also lives in the northern hemisphere and in forests. It builds a den to live in with its family while breeding. When it isn't breeding, it sleeps in the open and keeps itself warm by wrapping its tail around itself. It also has many ways to survive.

    The Pygmy Fowlx has powerful hearing from its big fox like ears that absorb the sound. They live up to three years in the wild, but can live up to ten years in captivity. Their large yellow eyes help them spot their prey. They also yelp to warn family members of danger. This is also a skill of communication and a way to show how they are social animals. This creature has twice as many bones in their neck than a human, which helps them see behind themselves. It also flies very fast to keep themselves away from predators.

    The Pygmy Fowlx has many predators like bears, wombats, mountain lions, eagles, grey wolves, coyotes, large mammals, and humans. They have many different predators and survival skills.

    The Pygmy Fowlx is a very fascinating and skillful creature with many different features. With many qualities, the Pygmy Fowlx can be social or whatever it wants to be. If you're hiking in the woods you might be lucky enough to see a Pygmy Fowlx flying.

www.ingramcontent.com/pod-product-compliance
Lightning Source LLC
Chambersburg PA
CBHW080247180526
45167CB00006B/2453